Introduction to Python and Large Language Models

A Guide to Language Models

Dilyan Grigorov

Apress®

Introduction to Python and Large Language Models: A Guide to Language Models

Dilyan Grigorov
Varna, Bulgaria

ISBN-13 (pbk): 979-8-8688-0539-4   ISBN-13 (electronic): 979-8-8688-0540-0
https://doi.org/10.1007/979-8-8688-0540-0

Copyright © 2024 by Dilyan Grigorov

This work is subject to copyright. All rights are reserved by the Publisher, whether the whole or part of the material is concerned, specifically the rights of translation, reprinting, reuse of illustrations, recitation, broadcasting, reproduction on microfilms or in any other physical way, and transmission or information storage and retrieval, electronic adaptation, computer software, or by similar or dissimilar methodology now known or hereafter developed.

Trademarked names, logos, and images may appear in this book. Rather than use a trademark symbol with every occurrence of a trademarked name, logo, or image we use the names, logos, and images only in an editorial fashion and to the benefit of the trademark owner, with no intention of infringement of the trademark.

The use in this publication of trade names, trademarks, service marks, and similar terms, even if they are not identified as such, is not to be taken as an expression of opinion as to whether or not they are subject to proprietary rights.

While the advice and information in this book are believed to be true and accurate at the date of publication, neither the authors nor the editors nor the publisher can accept any legal responsibility for any errors or omissions that may be made. The publisher makes no warranty, express or implied, with respect to the material contained herein.

> Managing Director, Apress Media LLC: Welmoed Spahr
> Acquisitions Editor: Celestin Suresh John
> Development Editor: James Markham
> Coordinating Editor: Gryffin Winkler

Cover designed by eStudioCalamar

Cover image by flatart on Freepik.com

Distributed to the book trade worldwide by Apress Media, LLC, 1 New York Plaza, New York, NY 10004, U.S.A. Phone 1-800-SPRINGER, fax (201) 348-4505, e-mail orders-ny@springer-sbm.com, or visit www.springeronline.com. Apress Media, LLC is a California LLC and the sole member (owner) is Springer Science + Business Media Finance Inc (SSBM Finance Inc). SSBM Finance Inc is a **Delaware** corporation.

For information on translations, please e-mail booktranslations@springernature.com; for reprint, paperback, or audio rights, please e-mail bookpermissions@springernature.com.

Apress titles may be purchased in bulk for academic, corporate, or promotional use. eBook versions and licenses are also available for most titles. For more information, reference our Print and eBook Bulk Sales web page at http://www.apress.com/bulk-sales.

Any source code or other supplementary material referenced by the author in this book is available to readers on GitHub (https://github.com/Apress). For more detailed information, please visit https://www.apress.com/gp/services/source-code.

If disposing of this product, please recycle the paper

*This book is dedicated to all my mentors at Google,
who not only sparked my passion for coding but also inspired
me to continuously pursue it without halt.*

Table of Contents

About the Author xv

About the Technical Reviewer xvii

Acknowledgments xix

Introduction xxi

Chapter 1: Evolution and Significance of Large Language Models 1

    The Evolutionary Steps of Large Language Models 2

    Markov, Shannon, and the Language Models 3

        Chomsky and the Language Models 5

    Rule-Based Language Models 6

    The First Chatbot: ELIZA 6

        Statistical Language Processing 8

        Vector Space Models and State Space Models 12

        Neural Language Models – The Rise of LLMs 13

        Attention-Based Language Models 20

        Large Language Models (LLMs) 22

        The Era of Multimodal Learning 22

    Understanding the NLP Basics 28

        What Exactly Is Natural Language Processing? 28

        How Does NLP Function? 29

        Elements of NLP 29

        NLP Tasks 31

        Text Preprocessing and Feature Engineering 32

        Word Embeddings and Semantic Understanding 43

        Sentiment Analysis and Text Classification with Python 47

        Exploring Approaches in Natural Language Processing (NLP) 52

TABLE OF CONTENTS

Delineating NLP Basics from LLM Capabilities 54
Contrasting Traditional NLP Techniques with LLMs 55
Summary 57

Chapter 2: What Are Large Language Models? 59

LM's Development Stages 59
How Do Large Language Models Work? 62
Overall Architecture of Large Language Models 63
In-Depth Architecture of the LLMs 64
    Tokenization 65
Attention 65
    Attention Mechanisms in LLMs 66
    Positional Encoding 67
    Activation Functions 68
    Layer Normalization 69
Architectures 72
    Encoder-Decoder 73
    Causal Decoder 73
    Prefix Decoder 73
    Pre-training Objectives 73
Model Adaptation 74
    Pre-training 74
    Alignment Verification and Utilization 75
    Prompting/Utilization 76
Training of LLMs 77
Benefits and Challenges of LLMs in Various Domains 78
    General Purpose 78
    Medical Applications 79
    Healthcare Communication and Management 80
    Enhanced Natural Language Processing 80
    Education 81
    Content Creation and Augmentation 82

Language Translation and Localization	82
Research and Data Analysis	82
Finance	82
Creative Arts	83
Ethical and Responsible Use	83
Legal and Compliance Assistance	83
Financial Analysis and Forecasting	84
Disaster Response and Management	84
Personalized Marketing and Customer Insights	84
Gaming and Interactive Entertainment	84
Accessibility Enhancements	85
Environmental Monitoring and Sustainability	85
LLMs and Engineering Applications	85
Chatbots	86
LLM Agents	87
LLM Limitations	87
Bias	88
Hallucinations	89
Beyond the Hype of the LLMs – Why Are They So Popular?	90
Common Benefits – The Real Reason Why LLMs Are So Popular	91
Large Language Models for Business	95
Summary	100
Chapter 3: Python for LLMs	**101**
Python at a Glance	101
Python Syntax and Semantics	102
Syntax Design Principles	103
Zen of Python	103
Python Identifiers	104
Python Indentation	105
Python Multiline Statements	106
Quotations in Python	106

TABLE OF CONTENTS

- Comments in Python 107
- How to Install Python and Your First Python Program 108
- Install Python on macOS 112
- Installing Python on Linux – Ubuntu/Debian and Fedora 113
- Your First Python Program 114
- Variables and Data Types, Numbers, Strings, and Casting 114
  - Naming a Variable 115
- Data Types 116
  - Numbers in Python 116
  - RAW Strings 120
  - Booleans and Operators 121
- Conditionals and Loops 132
  - Conditionals 132
  - Grouping Statements 133
  - Nested Blocks 134
  - Else and Elif Clauses 135
  - One-Line if Statements 136
- Python Loops (For and While) 136
  - While Loop in Python 137
  - Else Statement with while Loop 138
  - Creating an Infinite Loop with Python while Loop 138
  - For Loops in Python 139
  - Else Statement with for Loop 140
  - Nested Loops in Python 140
  - Loop Control Statements 142
- Python Data Structures: Lists, Sets, Tuples, Dictionaries 142
  - What Is a Data Structure? 143
  - Built-In Data Structures 144
  - Additional List Operations 147
  - Dictionaries in Python 148
  - Tuples in Python 152
  - Sets in Python 155

Regular Functions and Lambda Functions	158
What Is a Function in Python?	158
The return Statement	159
Return or Print in a Function	160
Methods vs. Functions	160
How to Call a Function in Python	160
Function Arguments in Python	161
Summary	164

Chapter 4: Python and Other Programming Approaches 165

Object-Oriented Programming in Python	165
Why Do We Use Object-Oriented Programming in Python?	166
Your First Python Object	168
Object-Oriented Programming (OOP) in Python Is Founded on Four Fundamental Concepts	169
Modules and File Handling	174
Python File Handling	180
The Powerful Features of Python 3.11	186
Understanding the Role of Python 3.11 in AI and NLP – Why Python?	195
Summary	197

Chapter 5: Basic Overview of the Components of the LLM Architectures 199

Embedding Layers	200
Stage 1: Nodes	202
Stage 2: Returning to the Words	203
Stage 3: Implementing the Softmax Layer	204
Feedforward Layers	205
Recurrent Layers	208
Attention Mechanisms	209
Understanding Tokens and Token Distributions and Predicting the Next Token	212
Understanding Tokenization in the Context of Large Language Models	212
The Advantages of Tokenization for LLMs	213
Limitations and Challenges	213

TABLE OF CONTENTS

Challenges in Current Tokenization Techniques 213
Tokenization Strategies in Large Language Models 215
Predicting the Next Token 218
Zero-Shot and Few-Shot Learning 222
Few-Shot Learning 222
Zero-Shot Learning 223
Few-Shot Learning 226
One-Shot Learning 226
Zero-Shot Learning 227
LLM Hallucinations 227
Classification of Hallucinations in Large Language Models (LLMs) 228
Implications of AI Hallucination 229
Mitigating the Risks of AI Hallucinations: Strategies for Prevention 230
When Hallucinations Might Be Good? 232
Future Implications 233
Examples of LLM Architectures 234
GPT-4 235
BERT 240
T5 245
Cohere 246
PaLM 2 247
Pathway Interaction and Collaboration 249
Selective Pathway Engagement 249
Generating Outputs 250
Jurassic-2 250
Claude v1 251
Falcon 40B 252
LLaMA 254
LaMDA 255
Guanaco-65B 257
Orca 258

TABLE OF CONTENTS

    StableLM 258

    Palmyra 259

    GPT4ALL 260

  Summary.. 261

Chapter 6: Applications of LLMs in Python 263

  Text Generation and Creative Writing 263

    The Mechanism Behind Text Generation 263

    The Significance of Text Generation 264

    Key Use Cases of Text Generation 264

    What Is Creative Writing 265

    Utilizing LLMs for Creative Writing Endeavors 265

    Blog Post Generator on a Topic and Length Provided by the User Based on OpenAI 266

  Language Translation and Multilingual LLMs 268

    Advantages of Utilizing LLMs for Translation 268

    How LLMs Translate Languages? 269

    Challenges Associated with LLMs in Translation 269

    The Potential Impacts of LLMs on the Translation and Localization Industry 270

  Text Summarization and Document Understanding................................... 273

    Article Summarization Application Using User-Provided URL 274

  Question-Answering Systems: Knowledge at Your Fingertips 279

    Enhancing Question-Answering Capabilities Through Large Language Models (LLMs) 279

    Utilizing Large Language Models for Advanced Document Analysis 279

    The Journey from Data to Response: A Comprehensive Overview 280

    Practical Applications and Use Cases of Generative Question Answering 281

    Question Answering Chatbot over Documents with Sources 282

    Full Code of the App 287

  Chatbots and Virtual Assistants 290

    What Is the Concept Behind Chatbots? 290

    Practical Applications of LLM-Trained Chatbots 290

    Guide to Building a Chatbot with LLMs.. 291

TABLE OF CONTENTS

 Customer Support Question Answering Chatbot 293
 Step 2: Formulate a Prompt for GPT-3 Utilizing Recommended Techniques 295
Basic Prompting – The Common Thing Between All Applications Presented........................... 299
 Understanding Prompting 299
 Fundamental Prompting Techniques 299
Summary 301

Chapter 7: Harnessing Python 3.11 and Python Libraries for LLM Development....303

LangChain 303
 LangChain Features 304
 What Are the Integrations of LangChain? 305
 How to Build Applications in LangChain? 305
 Use Cases of LangChain 306
 Example of a LangChain App – Article Summarizer 307
Hugging Face 309
 History of Hugging Face 310
 Key Components of Hugging Face 310
OpenAI API 316
 Features of the OpenAI API 317
 Industry Applications of the OpenAI API 318
 Simple Example of a Connection to the OpenAI API 320
Cohere 322
 Cohere Models 323
Pinecone 327
 How Vector Databases Operate 327
 What Exactly Is a Vector Database? 328
 Pinecone's Features 328
 Practical Applications 329
Lamini.ai 332
 Lamini's Operational Mechanics 332
 Lamini's Features, Functionalities, and Advantages 332
 Applications and Use Cases for Lamini 333

TABLE OF CONTENTS

Data Collection, Cleaning, and Preparation of Python Libraries 337
    Gathering and Preparing Data for Large Language Models 337
    Data Acquisition 338
    What Is Data Preprocessing? 338
    Preparing Datasets for Training 339
    Managing Unwanted Data 339
    Handling Document Length 343
    Text Produced by Machines 344
    Removing Duplicate Content 344
    Data Decontamination 345
    Addressing Toxicity and Bias 347
    Protecting Personally Identifiable Information (PII) 350
    Managing Missing Data 350
    Enhancing Datasets Through Augmentation 353
    Data Normalization 353
    Data Parsing 356
    Tokenization 358
    Stemming and Lemmatization 359

Feature Engineering for Large Language Models 362
    Word Embeddings 362
    Contextual Embeddings 362
    Subword Embeddings 363

Best Practices for Data Processing 364
    Implementing Strong Data Cleansing Protocols 365
    Proactive Bias Management 365
    Implementing Continuous Quality Control and Feedback Mechanisms 365
    Fostering Interdisciplinary Collaboration 365
    Prioritizing Educational Growth and Skill Development 366

Delving into Key Libraries 366

Summary 368

Index 369

About the Author

Dilyan Grigorov is a software developer with a passion for Python software development, generative deep learning and machine learning, data structures, and algorithms. He is an advocate for open source and the Python language itself. He has 16 years of industry experience programming in Python and has spent five of those years researching and testing generative AI solutions. Dilyan is a Stanford student in the Graduate Program on Artificial Intelligence in the classes of people like Andrew Ng, Fei-Fei Li, and Christopher Manning. He has been mentored by software engineers and AI experts from Google and Nvidia. His passion for AI and ML stems from his background as an SEO specialist dealing with search engine algorithms daily. He enjoys engaging with the software community, often giving talks at local meetups and larger conferences. In his spare time, he enjoys reading books, hiking in the mountains, taking long walks, playing with his son, and playing the piano.

About the Technical Reviewer

Tuhin Sharma is Sr. Principal Data Scientist at Red Hat in the Data Development Insights and Strategy group. Prior to that, he worked at Hypersonix as an AI architect. He also co-founded and has been CEO of Binaize, a website conversion intelligence product for e-commerce SMBs. He received a master's degree from IIT Roorkee and a bachelor's degree from IIEST Shibpur in Computer Science. He loves to code and collaborate on open source and research projects. He has four research papers and five patents in the field of AI and NLP. He is a reviewer of IEEE MASS conference in the AI track. He writes deep learning articles for O'Reilly in collaboration with the AWS MXNET team. He is a regular speaker at prominent AI conferences like O'Reilly Strata & AI, ODSC, GIDS, Devconf, etc.

Acknowledgments

As I reflect on the journey of writing this book, I am overwhelmed with gratitude for the countless individuals who have supported, inspired, and guided me along this path.

First and foremost, I extend my deepest thanks to my family, whose unwavering support and endless patience have been my anchor and source of strength.

I owe a great debt of gratitude to my mentors and colleagues, who have shared their wisdom, critiqued my ideas with kindness, and encouraged me to push the boundaries of my knowledge. Special thanks to my mentor Alexandre Blanchette whose insightful feedback and encouragement were invaluable to the completion of this manuscript.

Very special thanks to another of my mentors – Haiguang Li, one of the best machine learning and AI experts I have ever met. You have been my north star throughout this writing process. Your belief in me has been a gift beyond measure. Your generous sharing of knowledge, patience, and encouragement has not just shaped this book but has transformed me as a writer, software engineer, and individual. My sincerest thanks for your invaluable contribution.

My appreciation extends to the team at Apress and Springer Group, especially my editor, Celestine Suresh John, whose keen eye and creative vision have significantly enhanced this book. Thank you for your patience, guidance, and commitment to excellence.

A special word of thanks goes to the teams behind the development of large language models. To the researchers and engineers at organizations like OpenAI, Google Brain, the Google Research team, and others who have pushed the boundaries of what's possible with AI and machine learning, and who generously share their findings and tools with the world. Your work has not only inspired this book but also revolutionized how we think about human–computer interaction.

Finally, to you, the reader, embarking on this journey to explore Python and large language models: I hope this book serves as a valuable guide and inspires you to delve deeper into the transformative power of programming and AI. Thank you for your curiosity and your commitment to learning!

Introduction

In the evolving landscape of technology, where the boundaries between science fiction and reality blur, lies a transformative tool: the large language model (LLM). These models, sophisticated engines of artificial intelligence, have not only redefined our interaction with machines but have also opened new avenues for understanding human language. This book, structured into seven comprehensive chapters, serves as both a beacon and a bridge for those embarking on a journey through the intricate world of LLMs and their application using Python.

Chapter 1, "Evolution and Significance of Large Language Models," lays the foundation. It's here we start our journey, unraveling the complex yet fascinating world of natural language processing (NLP) and large language models. With over 50 pages dedicated to setting the stage, this chapter aims to provide the reader with a solid understanding of the evolution, significance, and basic concepts underpinning LLMs. Through a meticulous exploration of topics such as text preprocessing, word embeddings, and sentiment analysis, we uncover the magic and mechanics of LLMs and their impact across various domains.

Chapter 2, "What Are Large Language Models?", shifts the focus to the tools that make working with LLMs possible, with a particular emphasis on "Python and Why Python for LLMs?" It demystifies Python – a language synonymous with simplicity and power in the world of programming. From basic syntax to the nuanced features of Python 3.11, readers will gain the necessary knowledge to navigate the subsequent chapters and harness Python for their LLM endeavors.

In Chapter 3, "Python for LLMs," we plunge into the heart of LLMs, dissecting their components and understanding their workings. This chapter covers everything from embedding layers to attention mechanisms, providing insights into the technical makeup of models like GPT-4, BERT, and others. It's a chapter designed to equip readers with a profound understanding of how LLMs predict the next token, learn from few examples, and, occasionally, hallucinate.

Chapter 4, "Python and Other Programming Approaches," is a practical guide to leveraging Python for LLM development. Here, readers will familiarize themselves with essential Python libraries, frameworks, and platforms such as Hugging Face and OpenAI

INTRODUCTION

API, exploring their use in building applications powered by LLMs. With a focus on data preparation and a showcase of basic examples built with each framework, this chapter is a testament to Python's role in the democratization of AI.

Chapter 5, "Basic Overview of the Components of the LLM Architectures," demonstrates the versatility and potential of LLMs through practical Python applications. Readers will learn how to employ LLMs for tasks ranging from text generation to chatbots, each accompanied by step-by-step examples. This chapter not only highlights the capabilities of LLMs but also inspires readers to envision and create their own applications.

Chapter 6 of your document, titled "Applications of LLMs in Python," explores how Large Language Models (LLMs) are used in various domains, focusing on text generation and creative writing. It details how LLMs can generate human-like text using models like RNNs and transformers. The chapter covers key use cases, including content creation, chatbots, virtual assistants, and data augmentation. It also highlights how LLMs assist in creative writing tasks, brainstorming, dialogue crafting, world-building, and experimental literature. Additionally, the chapter discusses language translation, text summarization, and document understanding, emphasizing LLMs' impact on improving accuracy and efficiency in these areas. Finally, the chapter presents an example of building a question-answering chatbot using LLMs.

Chapter 7 explores how Python 3.11 and libraries such as LangChain, Hugging Face, and others are utilized to develop applications powered by large language models (LLMs). It covers the features of LangChain, such as model interaction, data connection, and memory, and explains how to build applications using these tools. The chapter also discusses the integrations and use cases of LangChain in various industries, like customer support, coding assistants, healthcare, and e-commerce, highlighting its flexibility in creating AI-powered solutions.

This book is an invitation to a world where understanding meets creation. It's for the curious minds eager to decode the language of AI and for the creators ready to shape the future. Whether you're a student, software engineer, data scientist, an AI or ML researcher or practitioner, or simply an enthusiast, the journey through these pages will equip you with the knowledge and skills to participate in the ongoing conversation between humans and machines. Welcome to the frontier of language, learning, and imagination.

CHAPTER 1

Evolution and Significance of Large Language Models

Over the recent decades, there have been notable advancements in language models and artificial intelligence technologies. Alongside advancements in computer vision, voice and speech processing, and image processing models, large language models (LLMs) are poised to profoundly impact the evolution of AI technologies. Therefore, it is crucial to examine the progress of language models since their inception and, more importantly, anticipate their future growth.

This chapter presents a concise overview of language models, tracing their development from statistical and rule-based models to today's transformer-based multimodal large language models with billions of parameters. It also aims to provide a good definition of what an LLM is and what are the mechanisms of work of the LLMs in general. Additionally, the benefits and limitations of existing models are observed and areas where current models require improvement are identified.

In recent times, substantial attention has been garnered by large language models, owing to numerous accomplishments in the field of natural language processing. Notably, the emergence of powerful chatbots like OpenAI's ChatGPT, Google Bard, and Meta's LLaMA, among others, has played a pivotal role. The achievements in language models are the culmination of decades of research and development. These models not only advance state-of-the-art NLP technologies but are also anticipated to significantly impact the evolution of AI technologies.

The foundational models, initially arising in NLP research, such as the transformers, have transcended into other domains like computer vision and speech recognition. However, AI models, especially language models, are not flawless, and the technology itself is akin to a double-edged sword. There are unresolved aspects that require further research and analysis. It remains uncertain whether these models are sophisticated computer programs generating responses solely through numerical computations and probabilities or if they possess a form of understanding and intelligence.

The sheer size of these gigantic language models makes it challenging to interpret their internal logic meaningfully. Hence, understanding the history of language models is crucial for depicting a better picture of their future development. The following paragraphs aim to provide a concise overview of language models and their evolution. Their goal is to review key milestones and innovations in language and machine learning models that have significantly influenced today's modern large language models.

That's a matter of subject that's in front of us to research in the future:

- Firstly, it offers insights into the core engineering elements of powerful language models, aiding in understanding the nature of different variations.

- Then it presents a taxonomy of related research, highlighting various categories of language models, their shortcomings, and how these issues have been addressed over time.

- The hope is that this overview will offer valuable insights into the past, present, and future of language models, contributing to the development of trustworthy models aligned with universal human values.

The Evolutionary Steps of Large Language Models

This section acknowledges the foundational work of Andrey Markov and Claude Shannon in the early 20th century, introducing concepts like n-grams and information theory that laid the groundwork for language modeling. It also reviews Noam Chomsky's introduction of the Chomsky hierarchy of grammars in 1956, which addressed the limitations of finite-state grammars and advocated for the use of context-free grammars in NLP.

Markov, Shannon, and the Language Models

The inception of language models traces back to the contributions of **Andrey Markov**[1] **and Claude Shannon**[2] in the early and mid-20th century, respectively. Markov laid the mathematical groundwork for what we now recognize as n-grams. Intriguingly, the concept of n-grams was first proposed in Markov's earlier works. Andrey Markov stands out as one of the earliest scientists to delve into the study of language models, even though the term "language model" had yet to be coined during that period.

Consider the conditional probability denoted as p(w1|w0) = p(w1). Various language models employ distinct approaches to compute conditional probabilities, denoted as p(wi|w1, w2, ..., wi-1). The procedure of acquiring and applying a language model is known as **language modeling**. An n-gram model serves as a fundamental type of model, assuming that the word at any given position is solely influenced by the words at the preceding n - 1 positions. In essence, the model operates as an **n - 1-order Markov chain**.

In 1906, Markov delved into the study of Markov chains, initially exploring a straightforward model with only two states and corresponding transition probabilities between them. He demonstrated that by traversing between these states based on the transition probabilities, the frequencies of transitioning to each state would ultimately converge to the expected values. Over subsequent years, Markov expanded the model and established that this conclusion held true in more generalized contexts.

Illustrating his model's practicality, Markov applied it to Alexander Pushkin's verse novel, "Eugene Onegin" in 1913. Through the removal of spaces and punctuation marks and the categorization of the first 20,000 Russian letters into vowels and consonants, Markov derived a sequence representing the novel's vowel and consonant distribution. Utilizing manual counting methods, Markov calculated the transition probabilities between vowels and consonants, using the resulting data to verify the characteristics of the simplest Markov chain.

---

[1] A. A. Markov, An Example of Statistical Investigation of the Text Eugene Onegin Concerning the Connection of Samples in Chains. www.alpha60.de/research/markov/,1913. 7., A. A. Markov, An example of statistical investigation in the text of `eugene onegin illustrating coupling `tests' in chains, Proceedings of the Academy of Sciences Of St.Petersburg, vol. 7,pp. 153–162, 1913

[2] C.E. Shannon, A mathematical theory of communication, The Bell System Technical Journal, vol. 27, no. 3, pp. 379–423, 1948

CHAPTER 1   EVOLUTION AND SIGNIFICANCE OF LARGE LANGUAGE MODELS

It is intriguing that the initial application area of the Markov chain was in language, with Markov's study serving as a foundational exploration into the **simplest form of a language model**.

Moving forward in the mid-20th century, **Claude Shannon** suggested the utilization of Markov processes to model natural language. Employing n-th order Markov chains, he developed a statistical model capable of characterizing the probability of sequences of letters, encompassing words and sentences. From a mathematical perspective, Shannon's approach involved counting the frequency of character sequences of length n, referred to as n-grams.

In 1948, Claude Shannon revolutionized the field of information theory with his seminal paper, "The Mathematical Theory of Communication." This work marked the inception of key concepts such as entropy and cross-entropy, as well as an exploration of the n-gram model. (Shannon adopted the term "entropy" from statistical mechanics following advice from John von Neumann.)

Entropy, in this context, signifies the uncertainty inherent in a probability distribution, while cross-entropy encapsulates the uncertainty of one probability distribution concerning another. Notably, entropy serves as a lower bound for cross-entropy. To elaborate, as the length of a word sequence tends toward infinity, the language's entropy can be defined, assuming a constant value that can be estimated from the language's data.

Shannon's contribution establishes a valuable tool for evaluating language models. If one language model demonstrates superior accuracy in predicting word sequences compared to another, it will exhibit lower cross-entropy. This insight provides a robust framework for assessing language modeling. It is essential to recognize that language models extend beyond natural languages to encompass formal and semi-formal languages.

The pioneering works of Markov and Shannon paved the way for diverse approaches to language modeling, encompassing rule-based systems, neural network–based systems, and more recently, **transformer-based pre-trained large language models** (LLMs) founded on attention mechanisms.

Chomsky and the Language Models

In 1956, Noam Chomsky[3] introduced the **Chomsky hierarchy of grammars**, a framework aimed at representing the **syntax of languages**. Chomsky emphasized the limitations of finite-state grammars, including n-gram models, in adequately describing natural languages.

Chomsky's theory posits that a language comprises a finite or infinite collection of sentences, where each sentence is a finite sequence of words drawn from a finite vocabulary. A grammar, according to Chomsky, is defined by a set of production rules capable of generating all sentences within the language. Different grammars yield languages of varying complexities, forming a **hierarchical structure**.

A grammar that can generate sentences acceptable by a finite-state machine is termed a finite-state grammar or regular grammar, whereas a grammar capable of producing sentences acceptable by a nondeterministic pushdown automaton is labeled a context-free grammar. Finite-state grammars are encompassed by context-free grammars.

The grammar underlying a finite Markov chain or an n-gram model aligns with a finite-state grammar. However, finite-state grammars face limitations in generating English sentences with grammatical relationships, such as those depicted in (i) and (ii).

- If S1, then S2.
- Either S3 or S4.
- Either if S5, then S6, or if S7, then S8.

While these relations can theoretically be combined indefinitely to create correct English expressions (as in example iii), finite-state grammars cannot account for all such combinations. Chomsky argued that there are significant constraints in describing languages, including n-gram models, with finite-state grammars. Instead, he advocated for the more effective modeling of languages using context-free grammars. This influence led to the increased adoption of context-free grammars in natural language processing (NLP) over the subsequent decades. Although Chomsky's theory may not wield significant influence in contemporary NLP, it retains important scientific value.

---

[3] N. Chomsky, Three models for the description of language. *IEEE Transactions on Information Theory 2*, 3 (1956), 113–124

Rule-Based Language Models

This section provides an overview of the first chatbot, ELIZA, created in 1966 by Joseph Weizenbaum. ELIZA's rule-based approach to dialogue simulation marked a significant milestone, although it lacked genuine comprehension of language nuances. This era saw a focus on rule-based chatbot programs, which led to the creation of several similar systems, including PARRY, ALICE, and CLEVER.

The First Chatbot: ELIZA

The exploration of language models traces back to the 1950s when researchers delved into **rule-based systems for language processing**. Initial attempts were hindered by the requirement to manually construct grammatical rules.

A significant breakthrough occurred in 1966 when Joseph Weizenbaum[4] introduced the "ELIZA" program (Figure 1-1). ELIZA stands out as one of the earliest computer programs capable of interacting with humans through chatting and question answering. Functioning as a simulated Rogerian psychotherapist, ELIZA engaged users in text-based conversations, representing a pioneering example of dialogue simulation. However, ELIZA's interactions relied heavily on predefined patterns and lacked genuine comprehension of the nuances in language.

---

[4] J. Weizenbaum, Eliza, a computer program for the study of natural language communication between man and machine, Communications of the ACM, vol. 9, no. 1, pp. 36–45, 1966

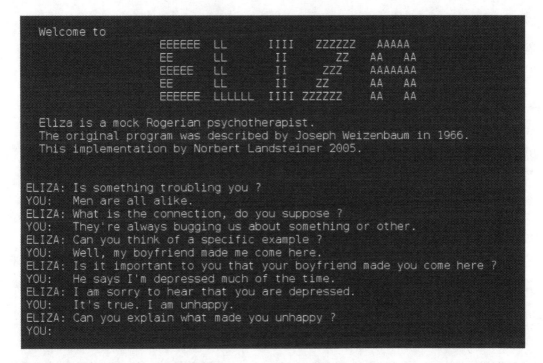

Figure 1-1. "ELIZA" program, Source: Wikipedia

Originating in 1966 at MIT and written in LISP-MAD, ELIZA utilizes a knowledge database to identify keywords with the highest rank when responding to a user's query. It then applies transformation rules to sentence patterns provided to the chatbot through scripts. An example of a popular script is called DOCTOR. ELIZA's success inspired researchers to expand on the idea and develop other rule-based chatbot systems.

For a period, there was a notable emphasis on rule-based chatbot programs among researchers, particularly following the development of ELIZA. Subsequently, several similar systems like **PARRY, ALICE, CLEVER, SHRDLU**, and others emerged in the ensuing years. These early rule-based chatbot systems were characterized by their simplicity and reliance on manually crafted rules and sentence patterns outlined in scripts provided to the system. **Consequently, the knowledge base and functionality of these programs were confined to the predefined rules and information.**

Over the subsequent decades, efforts to enhance language understanding saw advancements with the development of more sophisticated rule-based systems. Despite progress, these systems **grappled with the complexities of human language**, struggling to capture contextual subtleties and adapt to diverse linguistic expressions. Researchers increasingly recognized the necessity of transitioning from rigid rule-based approaches to models capable of learning and generalizing from data. This shift gave rise to statistical language processing techniques like n-grams and hidden Markov models, opening the door to more nuanced language analysis.

This distinction also highlights a significant difference between early rule-based chatbot programs and contemporary advanced models based on neural networks and large language models (LLMs). Another category of early language models relied on information retrieval (IR) techniques. In this approach, chatbots generated responses by matching patterns in pre-constructed databases of conversation pairs. Information retrieval–based chatbots, exemplified by CleverBot from the late 1980s, benefited from techniques like term frequency-inverse document frequency (TF-IDF), cosine similarity, and state space models at the word level.

Statistical Language Processing

While some early designs and models in the field of language modeling have become obsolete in today's advanced architectures, certain fundamental techniques have become integral components of contemporary language models and broader machine learning approaches. Key building blocks for natural language processing include n-grams, bag-of-words (BOW), and TF-IDF.

N-grams

The introduction of n-gram models in the 1990s and early 2000s marked a pivotal advancement in statistical language modeling. Founded on a simple yet potent concept, these models evaluated the likelihood of a word's occurrence by examining the **preceding words within a sequence**.

Despite their uncomplicated nature, n-gram models presented a crucial mechanism for grasping context in language. They represent sequences of n adjacent items from a sample of data, such as words in a sentence. By concentrating on the local relationships between words, these models began capturing the inherent dependencies that shaped meaningful linguistic expressions.

Despite their simplicity, n-grams have played a crucial role in **predicting the probability of the nth word based on the previous n-1 words**. This concept has been instrumental in both basic language modeling and the development of **more sophisticated models**.

> **Note** Google PageRank Algorithm
>
> *A noteworthy application of n-grams emerged with Google's groundbreaking PageRank[5] algorithm in 1996. This algorithm transformed web search by employing n-gram analysis to evaluate word co-occurrences across web pages, effectively ranking their relevance. This innovative application showcased the versatility of n-gram models in real-world scenarios extending beyond language processing alone. N-gram models not only underscored the importance of contextual information in language but also paved the way for the development of more intricate techniques capable of capturing a broader spectrum of linguistic nuances.*

Bag-of-Words (BOW)

Another fundamental technique in language modeling is the bag-of-words (BOW). This simple approach represents elements of language as numerical values based on word frequency in a document. Essentially, BOW[6] utilizes word frequency to create fixed-length vectors for document representation.

Although straightforward, BOW has been a foundational vectorization or embedding technique. Modern language models often build upon advanced word embedding and tokenization techniques.

The BOW model represents a method for converting a document into numerical form, a prerequisite before employing it in a machine learning algorithm. In any natural language processing task, this initial conversion is essential as machine learning algorithms cannot operate on raw text; thus, we must transform the text into a numerical representation, a process known as **text embedding**.

Text embedding involves two primary approaches: word vectors and document vectors. In the word vectors approach, each word in the text is represented as a vector (a sequence of numbers), and the entire document is then converted into a sequence of these word vectors. Conversely, document vectors embed the entire document as a single vector, simplifying the process compared to individual word embedding.

---

[5] Lawrence Page, S. Brin, The PageRank Citation Ranking: Bringing Order to the Web, 1996

[6] Z.S. Harris, Distributional structure, Word, vol. 10, no. 2–3, pp. 146–162, 1954

Moreover, it ensures that all documents are embedded in the same size, a convenience for machine learning algorithms that often require a fixed-size input. For instance, with a vocabulary of 1000 words, the document is expressed as a 1000-dimensional vector, where each entry signifies the frequency of the corresponding vocabulary word in the document.

While this technique may be limited for complex tasks, it serves well for simpler classification problems. Its simplicity and ease of use make it an appealing choice for embedding a set of documents and applying various machine learning algorithms. The BOW model offers easy implementation and swift execution.

Unlike other embedding methods that often demand specialized domain knowledge or extensive pre-training, this approach avoids such complexities or even manual feature engineering. It essentially works out of the box. However, its efficacy is limited to relatively simple tasks that **do not rely on understanding the contextual nuances of words**.

The typical application of the bag-of-words model is in embedding documents for classifier training. Classification tasks involve categorizing documents into multiple types, and the model's features are particularly effective for tasks like **spam filtering, sentiment analysis, and language identification**.

For instance, spam emails can be identified based on the frequency of key phrases like "act now" and "urgent reply," while sentiment analysis can discern positive or negative tones using terms like "boring" and "awful" vs. "beautiful" and "spectacular." Additionally, language identification becomes straightforward when examining the vocabulary.

Once documents are embedded, they can be fed into a classification algorithm. Common choices include the naive Bayes classifier, logistic regression, or decision trees/random forests – options that are relatively straightforward to implement and understand compared to more complex neural network solutions.

TF-IDF (Term Frequency-Inverse Document Frequency)

TF-IDF[7] is yet another statistical measure that leverages word frequency to evaluate the relevance of a word in a given set of documents. In comparison to bag-of-words, TF-IDF is more advanced as it employs two distinct metrics to quantify the relationship between words and documents in a more precise manner.

---

[7] H.P. Luhn, Statistical approach to mechanized encoding and searching of literary information, IBM Journal of Research and Development, vol. 1, no. 4, pp. 309–317, 1957

TF-IDF finds applications in information retrieval and text mining, particularly in tasks like searching for keywords in documents. It is also a valuable tool in natural language processing applications and language models. Term frequency-inverse document frequency (TF-IDF) also stands as a widely employed statistical method in gauging the significance of a term within a document relative to an entire document collection, known as a corpus.

In the process of text vectorization, words within a text document are converted into importance scores, and TF-IDF represents one of the most prevalent scoring schemes. Essentially, TF-IDF scores a word by multiplying its term frequency (TF) with the inverse document frequency (IDF).

Term frequency (TF) measures the frequency of a term within a document relative to the total number of words in that document.

$$TF = \frac{\text{number of times the term appears in the document}}{\text{total number of terms in the document}}$$

On the other hand, inverse document frequency (IDF) assesses the proportion of documents in the corpus that contain the term. Terms unique to a small percentage of documents, such as technical jargon terms, receive higher IDF values compared to common words found across all documents, like "a," "the," and "and."

$$IDF = \log\left(\frac{\text{number of the documents in the corpus}}{\text{number of documents in the corpus contain the term}}\right)$$

The TF-IDF score for a term is determined by the multiplication of its TF and IDF scores. In simpler terms, a term holds high importance when it appears frequently in a specific document but infrequently across others. This balance between commonality within a document, measured by TF, and rarity between documents, measured by IDF, results in the TF-IDF score, indicating the term's significance for a document in the corpus.

$$TF\text{-}IDF = TF * IDF$$

TF-IDF finds applications in various natural language processing tasks, including search engines, where it is used to rank document relevance for a query. It is also applied in text classification, text summarization, and topic modeling.

It is important to note that different approaches exist for calculating the IDF score. The calculation often involves using the base 10 logarithm, although some libraries may opt for a natural logarithm. Additionally, adding one to the denominator is a common practice to prevent division by zero.

Vector Space Models and State Space Models

Notably, the bag-of-words (BOW) model is considered a **basic word embedding technique**, correlating words to vectors. However, the widespread adoption of embedding words in vector spaces saw significant advancements, particularly with breakthroughs in the early 2000s and later with **Word2Vec**[8] **and GloVe in 2013 and 2014**, respectively. Word embedding techniques have become integral to language modeling, extensively employed by various other models.

Vector space models and state space models are fundamental concepts in language modeling. Vector space models involve **algebraic approaches** for representing language elements as vectors embedded in specific vector spaces.

A vector space comprises vectors, numerical representations of words, sentences, and even documents. While basic vectors, such as map coordinates, have only two dimensions, those employed in natural language processing can encompass thousands.

The concept of representing words as vectors, often known as **word embedding or word vectorization**, has been present since the 1980s.[9] This simplifies the assessment of similarity between words or the relevance of a search query to a document. Cosine similarity is commonly applied to gauge the similarity between vectors. Utilizing linear algebra with nonbinary term weights, the vector space model enables the computation of the continuous degree of similarity between two objects, such as a query and documents, facilitating partial matching.

---

[8] T. Mikolov, K. Chen, G. Corrado, and J. Dean, Efficient estimation of word representations in vector space, preprint arXiv:1301.3781,2013

[9] D.E. Rumelhart, G.E. Hinton, R.J. Williams, et al., Learning internal representations by error propagation, 1985

> **Note** State Space Models
>
> *Although vector space models have been thoroughly examined and utilized in natural language processing (NLP), there has been a recent surge of interest in state space models within the NLP domain. Despite this, state space models have already demonstrated their value in sequence modeling across various domains, including signal processing and control theory. Acknowledging the effectiveness of state space models in managing lengthy sequences, these models offer promising possibilities for crafting innovative architectures in language modeling and refining existing techniques.*

Neural Language Models – The Rise of LLMs

In 2001, Yoshua Bengio and colleagues introduced one of the earliest neural language models, marking the onset of a new era in language modeling. It's noteworthy that Bengio,[10] Geoffrey Hinton, and Yann LeCun were honored with the 2018 ACM A.M. Turing Award for their groundbreaking contributions, widely recognized for making deep neural networks a **vital component of computing**.

The conventional n-gram model exhibits limitations in its learning capacity. The customary approach involves estimating conditional probabilities, p(wi|wi-n+1, wi-n+2, ..., wi-1), from the corpus using a smoothing method. However, as the value of n increases, the model's parameters grow exponentially at O(Vn), where V represents the vocabulary size. This exponential growth, coupled with the sparsity of training data, hinders the accurate learning of model parameters.

Bengio et al.'s neural language model enhances the n-gram model through two key advancements. Firstly, it incorporates a real-valued vector known as word embedding to represent individual words or word combinations. Compared to the "one-hot vector" representation of a word, where the corresponding element is one and others are zero, word embedding has significantly lower dimensionality.

---

[10] Y. Bengio, R. Ducharme, and P. Vincent, A neural probabilistic language model. In Advances in Neural Information Processing Systems (2001), 932–938

Functioning as a form of "distributed representation," word embedding offers enhanced efficiency, generalization ability, robustness, and extensibility compared to the one-hot vector. Secondly, the language model is structured as a neural network, leading to a substantial reduction in the number of parameters within the model.

Artificial neural networks (ANNs) have emerged as a widely adopted approach in natural language processing and the development of various machine learning and artificial intelligence models.

Following the contributions of Bengio and colleagues, a plethora of techniques for word embedding and neural language modeling have emerged, introducing enhancements from various angles.

The late 2000s witnessed a pivotal shift marked by a renewed fascination with neural networks, especially the emergence of deep learning techniques. The utilization of **recurrent neural networks (RNNs) and Long Short-Term Memory (LSTM) networks**[11] for processing sequential data played a crucial role in advancing language understanding and generation. This transformative period saw notable applications, particularly in machine translation. Google's introduction of the "Neural Machine Translation" system in 2016 exemplified this progress, outperforming traditional methods by leveraging deep learning techniques for language translation.

A notable milestone during this era was the unveiling of the Word2Vec model in 2013 by **Tomas Mikolov** and his team at Google. This model revolutionized language representation by learning distributed word representations based on their contextual usage within extensive text datasets. The Word2Vec model laid the groundwork for **capturing semantic relationships between words**, elevating language understanding beyond mere statistical associations.

Recurrent Neural Networks

Recurrent neural networks (RNNs) (Figure 1-2) represent a pivotal advancement in language models following basic statistical and rule-based models. Originating in the 1980s, RNNs have played a significant role in processing sequential and time series data. While contributing to language model progress, RNNs face challenges in handling

---

[11] S. Hochreiter and J. Schmidhuber, Long short-term memory. *Neural Computation* 9, 8 (1997), 1735–1780

long-term dependencies,[12] such as the issues of vanishing gradient and exploding gradient. Researchers have sought improved solutions to address these limitations and enhance the capability of processing lengthy sequences.

A crucial aspect within the realm of recurrent neural networks (RNNs) is the concept of intermediate representations or states. The interdependence among words is mirrored by the relationships between states in the RNN model. While the model's parameters are shared across different positions, the resulting representations differ based on their respective positions.

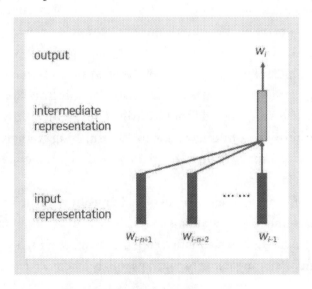

Figure 1-2. *Basic architecture of the layers of RNN*

In the RNNs at each position, there exists an intermediate representation for each layer, embodying the "state" of the word sequence up to that point. The intermediate representation for the current layer at the current position is determined by the intermediate representation of the same layer at the previous position and the intermediate representation of the layer beneath at the current position. The ultimate intermediate representation at the current position plays a pivotal role in computing the probability of the next word.

---

[12] Y. Bengio, P. Simard, and P. Frasconi, Learning long-term dependencies with gradient descent is difficult, IEEE Transactions on Neural Networks, vol. 5, no. 2, pp. 157–166, 1994

Long Short-Term Memory

Long Short-Term Memory (LSTM) networks represent a type of deep learning sequential neural network designed to address the challenge of information retention. Functioning as a specialized form of recurrent neural network (RNN), LSTMs effectively tackle the vanishing gradient problem encountered by traditional RNNs, a significant issue in machine learning algorithms. **Hochreiter and Schmidhuber**[13] **developed LSTM to overcome limitations inherent in conventional RNNs**, presenting a solution that enhances the performance of machine learning models. Implementation of the LSTM model can be carried out using the Keras library in Python.

To illustrate, consider the experience of recalling a prior scene while watching a video or passage of a music song, or just remembering events from an earlier chapter while reading a book. Analogously, RNNs operate by retaining previous information and utilizing it for processing current input. However, RNNs **face a limitation in remembering long-term dependencies** due to the vanishing gradient problem. LSTMs, on the other hand, are expressly crafted to circumvent these challenges associated with long-term dependencies.

Differing from conventional neural networks, Long Short-Term Memory (LSTM) integrates feedback connections, enabling it to process entire sequences of data rather than individual data points. This distinctive capability makes it highly proficient in comprehending and predicting patterns within sequential data, including time series, text, and speech.

In the realm of artificial intelligence and deep learning, LSTM has emerged as a potent tool, contributing to breakthroughs across various fields by extracting valuable insights from sequential data.

At a broad level, the functioning of LSTM closely resembles that of an RNN cell. The internal mechanism of the LSTM network involves three key components, each with a distinct function:

- Forget Gate
- Input Gate
- Output Gate

---

[13] S. Hochreiter and J. Schmidhuber, Long short-term memory, Neural computation, vol. 9, no. 8, pp. 1735–1780, 1997

The initial part decides whether information from the previous timestamp is relevant and should be retained or deemed irrelevant and discarded. The second segment focuses on learning new information from the input. Finally, in the third section, the updated information from the current timestamp is passed to the subsequent timestamp, constituting a single time step in the LSTM cycle.

These three components are also known as **gates**. The gates govern the flow of information in and out of the memory cell or LSTM cell. The Forget Gate, the Input Gate, and the Output Gate regulate these flows. An LSTM unit, comprising these gates along with a memory cell or LSTM cell, can be likened to a layer of neurons in a **traditional feedforward neural network**, where each neuron encompasses a hidden layer and a current state.

Similar to a basic RNN, LSTM possesses a hidden state denoted as H(t-1) for the previous timestamp and H(t) for the current timestamp. Additionally, LSTM incorporates a cell state represented by C(t-1) and C(t) for the previous and current timestamps, respectively. The hidden state is identified as **Short-Term Memory**, while the cell state is termed **Long-Term Memory**.

Example: *Consider two sentences separated by a full stop: "John is a nice person," followed by "Allison, on the other hand, is bad." As we transition from the first sentence to the second, the LSTM network, utilizing the Forget Gate, can appropriately disregard information about John and acknowledge that the focus has shifted to Allison.*

Note Bidirectional LSTMs

*Moving beyond LSTM, bidirectional LSTMs represent a type of recurrent neural network (RNN) architecture that processes input data in both forward and backward directions. Unlike traditional LSTMs that consider only past-to-future information flow, bidirectional LSTMs incorporate **future-to-past context**, enhancing their ability to capture dependencies in both directions.*

*Comprising two LSTM layers—one for forward processing and the other for backward processing—this architecture proves particularly effective in tasks requiring a comprehensive understanding of input sequences, such as **sentiment analysis, machine translation, and named entity recognition**. The bidirectional LSTMs' incorporation of information from both directions enhances their capacity to capture **long-term dependencies**, leading to more accurate predictions in complex sequential data.*

Gated Recurrent Units

Gated Recurrent Units (GRUs), introduced by Cho et al. in 2014,[14] present another approach to mitigate the vanishing gradient problem in RNNs. GRUs share similarities with LSTM networks by incorporating additional gates to handle the gradient challenge. However, GRUs utilize only two gates (**update and reset**) and lack an extra cell present in LSTMs. Consequently, GRUs boast a simpler architecture with fewer parameters, resulting in faster training. Nevertheless, LSTMs and GRUs each have distinct strengths and weaknesses, making them suitable for different applications. Notably, **LSTM networks may excel in scenarios with long-term dependencies in the data**.

Sequence-to-Sequence Language Models

A significant advancement in the development of language models is marked by the inception of sequence-to-sequence (*seq2seq*) models. The primary impetus behind this family of language models lies in addressing machine translation challenges. Tasks related to machine translation commonly involve solving **sequence-to-sequence or sequence modeling problems**. In these scenarios, the objective is to create models capable of mapping a sequence from one domain to a corresponding sequence in another domain, exemplified by translating a sentence from English to Italian, for example.

A prevalent strategy in designing seq2seq models[15] involves employing **encoder-decoder architectures**, where both the encoder and decoder are based on **recurrent neural network (RNN) architectures like LSTM or GRU**. The encoder processes the input sequence, **generating a vector representation, or context vector, of the input**. The decoder then utilizes this context vector to generate the output sequence.

Although Seq2Seq models gained popularity in machine translation tasks, they exhibited limitations, primarily due to their reliance on RNN-based architectures such as LSTMs or GRUs. A notable drawback of Seq2Seq architectures was their **inability to discern the importance of different parts within a sequence**. In other words, they processed a sequence without recognizing whether specific segments held more significance than others.

---

[14] K. Cho, B. Van Merriënboer, C. Gulcehre, D. Bahdanau, F. Bougares, H. Schwenk, and Y. Bengio, Learning phrase representations using RNN encoder-decoder for statistical machine translation, arXivpreprintarXiv:1406.1078, 2014

[15] I. Sutskever, O. Vinyals, and Q.V. Le, Sequence to sequence learning with neural networks, Advances in neural information processing systems, vol. 27, 2014

This deficiency prompted researchers to explore methods for enabling models to selectively focus on crucial parts of a sequence. This exploration gave rise to attention-based language models and, subsequently, the transformer architecture.

Note Conditional Language Model

Language models serve a dual purpose: they can compute the probability of a language (word sequence) or generate language itself. In the latter scenario, natural language sentences or articles are produced, often through random sampling from language models. It's well established that LSTM language models, trained on substantial datasets, exhibit the capability to generate highly natural sentences.

An extension of the language model concept is the conditional language model, which assesses the conditional probability of a word sequence given a specific condition. When this condition is another word sequence, the task transforms into what is commonly known as the sequence-to-sequence problem. This problem is evident in tasks like machine translation, text summarization, and generative dialogue. Alternatively, if the condition is an image, the challenge becomes transforming an image into a word sequence, exemplified by tasks like image captioning.

Conditional language models find applications across a diverse range of tasks:

- In machine translation, these systems convert sentences from one language to another while preserving the semantics.

- In dialogue generation, the system crafts a response to the user's input, constituting one round of dialogue.

- Text summarization involves transforming a lengthy text into a concise representation that encapsulates the essence of the original content.

The semantics embodied in the conditional probability distributions of these models vary based on the specific application and are acquired through data-driven learning within each application context.

Attention-Based Language Models

Attention-based language models are a type of neural network architecture designed to handle sequential data by focusing on the most relevant parts of the input for each output. They use attention mechanisms like self-attention to weigh the importance of different words in a sentence, enabling them to capture long-range dependencies and context more effectively than traditional models. This approach has significantly improved performance in natural language processing tasks, such as translation and text generation.

Attention Mechanism

For decades, researchers have endeavored to develop artificial systems and integrate elements of the human visual system and brain mechanisms into these systems. A significant stride toward achieving this objective occurred in 2014, where researchers introduced a fundamental yet ingenious concept: **the incorporation of an attention mechanism**[16] **into encoder-decoder architectures**.

The key proposition was that the relative importance of elements within a sequence could be encoded and assigned during sequence processing. The attention technique proposed an alignment model to compute scores, which were then **subjected to a softmax function to generate weights**. Subsequently, a context vector was derived through a weighted sum of the encoder's hidden states.

The integration of the attention mechanism stands as a pivotal concept in advancing artificial neural networks and, more broadly, machine learning models. Notably, it served as inspiration for several influential breakthroughs in natural language processing (NLP), including the **renowned transformer architecture**. This transformer architecture, in turn, paved the way for various innovative developments such as the **vision transformer (ViT), the BERT model, and GPT models**, leading to the creation of powerful chatbot programs like ChatGPT.

---

[16] D. Soydaner, Attention mechanism in neural networks: where it comes and where it goes, Neural Computing and Applications, vol. 34, no. 16, pp. 13371–13385, 2022

The Transformer Architecture

A pivotal development in the evolution of language models is the introduction of the transformer architecture in 2017. Drawing inspiration from the encoder-decoder architecture, the transformer architecture derives its potency from the attention mechanism, alongside other components like word embedding and positional encoding.

The principal distinction between the transformer architecture and earlier RNN-based architectures lies in its nonsequential data processing approach. Notably, the transformer cleverly combines word embedding and positional encoding techniques, facilitating parallel data processing. This attribute contributes to the transformer's effectiveness, enabling benefits from parallelization and expediting training.

Prior to 2017, recurrent networks, particularly Long Short-Term Memory (LSTM), stood as the prevalent architecture for language modeling. LSTMs boasted **theoretically infinite context size**, incorporating both past and future words through a bidirectional approach. Despite the theoretical advantages, the practical benefits were limited, and the recurrent structure incurred higher training costs and time consumption, particularly challenging for parallelization on GPUs. It was primarily due to these challenges that **transformers emerged as a replacement for LSTMs**.

Transformers leverage the attention mechanism, where the model learns to assign weights to words based on contextual relevance. In contrast to recurrent models, where the most recent word has the most immediate impact on predicting the next word, attention allows all words in the current context to be accessible, empowering the model to discern which words to prioritize.

In their aptly titled paper, "Attention is All You Need,"[17] **Google researchers introduced the Transformer sequence-to-sequence architecture**. This revolutionary model abandons recurrent connections, utilizing its own output as context during text generation. This design choice facilitates easy parallelization of training, enabling the scaling of models and training data to unprecedented sizes.

For classification tasks, the **Bidirectional Encoder Representations from Transformers**[18] **(BERT)** emerged as the preferred model. In the realm of text generation, the focus shifted toward scaling up models.

---

[17] https://research.google/pubs/attention-is-all-you-need/
[18] J. Devlin, M.-W. Chang, K. Lee, and K. Toutanova, Bert: Pre-training of deep bidirectional transformers for language understanding, arXiv preprint arXiv:1810.04805, 2018

Widely recognized as a foundational model for artificial intelligence (AI), the transformer architecture has exerted substantial influence across diverse domains of artificial intelligence, including natural language understanding (e.g., BERT), computer vision (e.g., ViT), and speech recognition.

In essence, the transformer architecture serves as a versatile neural computer applicable in various downstream NLP technologies and multimodal AI applications. It is particularly well suited for deployment in multimodal large language models capable of processing diverse data types such as text, images, and videos.

Large Language Models (LLMs)

The culmination of breakthroughs in artificial neural networks and advancements in natural language processing models has given rise to a variety of large language models (LLMs), showcasing remarkable capabilities in NLP tasks. Prominent examples of these LLMs include **OpenAI's ChatGPT and GPT-4, Google's Bard, Meta's LLaMA, Anthropic's Claude, and BloombergGPT**.

The potency and success of most large language models can be attributed to preceding innovations in the evolution of language models. The transformer architecture, built upon the encoder-decoder model and incorporating the attention mechanism, represents a significant advancement that has paved the way for powerful large language models.

Additional contributing factors include the utilization of extensive datasets for model training. While LLMs have demonstrated significant efficacy in specific NLP tasks such as question answering and text summarization, there remain various aspects in which LLMs need further improvement.

The Era of Multimodal Learning

Multimodal deep learning refers to the exploration of computer algorithms that enhance their performance by learning from diverse datasets that incorporate multiple modalities.

In the realm of machine learning, multimodal deep learning is a specialized subfield focused on training artificial intelligence models to analyze and **establish connections among various types of data, known as modalities**. These modalities often include images, video, audio, and text. Through the integration of different modalities, a deep learning model can gain a more comprehensive understanding of its surroundings, as certain cues are exclusive to specific modalities.

To illustrate, consider the task of emotion recognition; while visual cues from a human face (visual modality) are crucial, the tone and pitch of a person's voice (audio modality) carry significant information about their emotional state, which may not be evident from facial expressions alone, even when they are synchronized.

While Unimodal or Monomodal models, which process only a single modality, have undergone extensive research and yielded impressive results in areas such as computer vision and natural language processing, they possess inherent limitations. As a result, there is a growing necessity for multimodal models.

Multimodal models frequently leverage deep neural networks, although earlier research has integrated other machine learning models like hidden Markov models (HMM) or Restricted Boltzmann Machines (RBM).

In the domain of multimodal deep learning, the prevalent modalities typically encompass visual elements (such as images and videos), textual information, and auditory components (including voice, sounds, and music). Nonetheless, less conventional modalities also play a role, such as 3D visual data, depth sensor data, and LiDAR data, particularly in applications like self-driving cars.

Nevertheless, the most prevalent combinations involve the four primary modalities:

- Image + Text
- Image + Audio
- Image + Text + Audio
- Text + Audio

Multimodal Learning Challenges

Multimodal deep learning is dedicated to addressing five fundamental challenges that currently serve as active research domains. The resolution or enhancement of any of these challenges holds the potential to propel forward the field of multimodal AI research and application.

Representation, the first challenge, involves the encoding of data from diverse modalities into a vector or tensor. The creation of effective representations that encapsulate the semantic information of raw data is crucial for the success of machine learning models. However, the difficulty lies in extracting features from heterogeneous data in a manner that optimally exploits their synergies. It is equally essential to fully harness complementarity among different modalities while avoiding the inclusion of redundant information.

Multimodal representations can be categorized into two types:

1. **Joint Representation:** Each individual modality undergoes encoding and is subsequently placed into a shared high-dimensional space. This direct approach may be effective, especially when modalities exhibit similar natures.

2. **Coordinated Representation:** Individual modalities are independently encoded, but their representations are coordinated through the imposition of restrictions. For instance, their linear projections should exhibit maximal correlation. This approach facilitates the integration of disparate modalities while maintaining coordination.

Fusion

The fusion task involves integrating information from two or more modalities to perform a prediction. Effectively merging diverse modalities, such as video, speech, and text, poses a challenge due to the heterogeneous nature of multimodal data.

Fusing heterogeneous information constitutes a fundamental aspect of multimodal research, accompanied by a myriad of challenges. Practical issues include addressing variations in formats, lengths, and non-synchronized data. Theoretical challenges revolve around identifying the optimal fusion technique, encompassing simple operations like concatenation or weighted sums, as well as more intricate mechanisms such as transformer networks or attention-based recurrent neural networks (RNNs).

The choice between early and late fusion is also a consideration. Early fusion integrates features immediately after extraction, employing mechanisms like those mentioned previously. In contrast, late fusion performs integration only after each unimodal network produces a prediction, often employing voting schemes, weighted averages, or other techniques. Hybrid fusion techniques, which combine outputs from early fusion and unimodal predictors, have also been proposed.

Alignment

Alignment involves identifying direct relationships between different modalities. Contemporary multimodal learning research strives to create modality-invariant representations, meaning that when different modalities signify a similar semantic concept, their representations should be close together in a latent space. For instance, the sentence "she dived into the pool," an image of a pool, and the audio signal of a splash sound should share proximity in the representation space manifold.

Translation

Translation entails mapping one modality to another, exploring how one modality (e.g., textual) can be translated to another (e.g., visual) while preserving semantic meaning. Translations are inherently open-ended and subjective, lacking a definitive answer, adding complexity to the task.

Current multimodal learning research involves constructing generative models for modal translations. Recent examples include DALL-E and other text-to-image models, illustrating the development of generative models facilitating translations between text and visual modalities.

Co-learning

Multimodal co-learning is directed toward transferring knowledge acquired from one or more modalities to tasks involving other modalities. This approach is particularly crucial in scenarios characterized by low-resource target tasks, the absence or partial presence of modalities, or the presence of noisy data.

Translation can be employed as a method of co-learning to facilitate the transfer of knowledge from one modality to another. Neuroscientific insights also indicate that humans may engage in co-learning through translation methods. Individuals experiencing aphantasia, the inability to mentally create images, demonstrate poorer performance on memory tests. Conversely, those adept at creating such mappings, such as translating between textual/auditory and visual modalities, exhibit enhanced performance on memory tests. This underscores the significance of the ability to convert representations across different modalities as a key aspect of human cognition and memory.

How Multimodal Learning Operates

Multimodal neural networks typically amalgamate multiple unimodal neural networks. For instance, in an audiovisual model, there are two distinct unimodal networks: one dedicated to processing visual data and another to audio data. Each of these unimodal networks independently processes its inputs through a step known as encoding. Subsequent to unimodal encoding, the extracted information from each model needs to be effectively fused together.

Various fusion techniques have been proposed, ranging from straightforward concatenation to the incorporation of attention mechanisms. The process of fusing multimodal data stands out as a critical determinant of success. Following the fusion process, a conclusive "decision" network receives the merged encoded information and undergoes training for the ultimate task.

In essence, multimodal architectures typically comprise three integral components:

1. Unimodal encoders that individually encode distinct modalities, usually with one encoder dedicated to each input modality

2. A fusion network responsible for amalgamating features extracted from each input modality during the encoding phase

3. A classifier that receives the fused data and generates predictions

These components are commonly referred to as the encoding module, the fusion module, and the classification module.

Applications of Multimodal Deep Learning

Multimodal deep learning, which integrates and processes information from various data sources such as text, images, and audio, is revolutionizing numerous fields. By harnessing the power of different modalities, this approach enhances the capabilities of AI systems, leading to more accurate and robust solutions. In this section, we will explore how LLM technology is being applied across different industries and the significant impact it is making.

Image Captioning

Image captioning involves generating brief textual descriptions for given images, utilizing multimodal datasets containing both images and text descriptions. It addresses the translation challenge by converting visual representations into textual ones. Models need to grasp the semantics of an image, detecting key objects, actions, and characteristics. This task can also extend to video captioning.

Image captioning models find utility in providing text alternatives to images, benefiting blind and visually impaired users.

Image Retrieval

Image retrieval entails finding relevant images within a large database based on a retrieval key. This task, also known as Content-Based Image Research (CBIR), leverages deep learning multimodal models, offering a broader solution with enhanced capabilities, often reducing the reliance on traditional tags. Image retrieval is evident in everyday scenarios, such as search engine results displaying images related to a query.

Text-to-Image Generation

Text-to-image generation is a popular multimodal learning application that tackles the translation challenge. Models like OpenAI's DALL-E and Google's Imagen create images based on short text descriptions, providing a visual representation of the text's semantic meaning. These models find application in photoshopping, graphic design, and digital art inspiration.

Emotion Recognition

Emotion recognition, a use case where multimodal datasets outperform monomodal ones, involves analyzing video, text, and audio data to identify emotional states. Incorporating sensor data like encephalogram data further enhances the multimodal input. While multimodal datasets convey more information, training multimodal networks poses challenges, sometimes leading to performance degradation compared to single modality counterparts. Understanding these difficulties is crucial for effective implementation.

In summary, multimodal deep learning finds diverse applications, enhancing the capabilities of computer vision systems across various tasks.

CHAPTER 1   EVOLUTION AND SIGNIFICANCE OF LARGE LANGUAGE MODELS

Understanding the NLP Basics

Natural language processing (NLP) pertains to a field within computer science, specifically falling under the broader umbrella of artificial intelligence (AI), which aims to equip computers with the capability to comprehend both written text and spoken language in a manner akin to human comprehension.

NLP merges elements of computational linguistics, involving rule-based modeling of human language, with statistical, machine learning, and deep learning models. These combined technologies empower computers to process human language, be it in the form of text or spoken communication, and to grasp its full meaning, encompassing the intentions and sentiments of the speaker or writer.

NLP drives the development of computer programs capable of tasks such as translating text from one language to another, responding to voice commands, and rapidly summarizing extensive textual content – even in real-time scenarios. Chances are high that you have interacted with NLP through voice-operated GPS systems, digital assistants, speech-to-text transcription software, customer service chatbots, and other everyday conveniences. Nevertheless, NLP also assumes an increasingly significant role in enterprise solutions, aiding in the optimization of business operations, enhancement of employee productivity, and simplification of critical business processes.

Natural language processing (NLP) holds a central position within the field of artificial intelligence, serving as a vital link between human communication and computer comprehension. It represents an interdisciplinary domain that grants computers the ability to comprehend, analyze, and generate human language, facilitating seamless interaction between humans and machines. The significance of NLP is readily apparent in its wide-ranging applications, spanning from automated customer support to real-time language translation.

What Exactly Is Natural Language Processing?

Natural language processing (NLP) emerges as a branch of artificial intelligence squarely focused on the interface between computers and humans employing natural language. Its primary aim is to instruct computers to process and scrutinize vast volumes of natural language data.

NLP encompasses the task of enabling machines to grasp, decipher, and generate human language in a manner that is both valuable and meaningful. OpenAI, renowned for developing advanced language models such as ChatGPT, underscores the pivotal role of NLP in crafting intelligent systems capable of comprehending, responding to, and generating text, thereby enhancing the user-friendliness and accessibility of technology.

How Does NLP Function?

NLP empowers computers to comprehend natural language in a manner akin to human understanding. Be it spoken or written language, natural language processing employs artificial intelligence to receive real-world input, process it, and extract meaning in a format comprehensible to computers. Much like humans possess various sensory organs, such as ears for hearing and eyes for seeing, computers rely on programs to read and microphones to capture audio data. Furthermore, just as humans employ their brains to process sensory input, computers utilize specific programs to process the data obtained through their respective input methods. At a certain stage of this processing, the input is transformed into code that the computer can interpret.

The natural language processing journey primarily consists of two key phases: data preprocessing and algorithm development.

Data preprocessing entails the preparation and refinement of textual data to make it amenable for machine analysis. This preprocessing step refines the data, rendering it in a format suitable for machine processing, and brings to light distinctive textual features that algorithms can effectively work with.

Elements of NLP

Natural language processing is not a single, unified approach but instead consists of several constituent elements, each contributing to the comprehensive comprehension of language. The primary components that NLP endeavors to fathom encompass syntax, semantics, pragmatics, and discourse.

Syntax

Syntax encompasses the organization of words and phrases to construct well-structured sentences within a language. In the sentence "The boy played with the ball" syntax entails scrutinizing the grammatical structure of this sentence to ensure it adheres to English grammatical rules, including subject-verb agreement and correct word order.

Semantics

Semantics is concerned with grasping the meanings of words and how they collectively convey meaning when used in sentences. In the sentence "The panda eats shoots and leaves," semantics assists in discerning whether the panda consumes plants (shoots and leaves) or engages in a violent act (shoots) and subsequently departs (leaves), depending on word meanings and context.

Pragmatics

Pragmatics deals with comprehending language within diverse contexts, ensuring that the intended meaning is derived based on the situation, the speaker's intent, and shared knowledge. When someone says, "Can you pass the salt?" pragmatics involves recognizing that it is a request rather than an inquiry about one's capability to pass the salt, interpreting the speaker's intent within the dining context.

Discourse

Discourse concentrates on the analysis and interpretation of language beyond the sentence level, considering how sentences interrelate within texts and conversations. In a conversation where one person exclaims, "I'm freezing," and another responds, "I'll close the window," discourse entails understanding the coherence between the two statements, acknowledging that the second statement is a response to the implied request in the first.

Comprehending these constituent elements is imperative for individuals venturing into NLP, as they constitute the foundation of how NLP models decipher and generate human language.

NLP Tasks

Human language is rife with complexities that pose significant challenges when it comes to developing software capable of accurately discerning the intended meaning from text or spoken data. Homonyms, homophones, sarcasm, idioms, metaphors, exceptions in grammar and usage, and variations in sentence structures are just a few of the intricacies of human language. While humans may take years to master these nuances, programmers must instill the ability to recognize and understand them accurately from the outset in natural language-driven applications, if these applications are to be truly effective.

Various NLP tasks are employed to deconstruct human text and voice data in ways that facilitate the computer's comprehension. Some of these tasks encompass:

- **Speech recognition**, also known as speech-to-text, involves the reliable conversion of voice data into textual form. Speech recognition is a fundamental requirement for applications that rely on voice commands or provide responses to spoken queries. What adds to the challenge of speech recognition is the way people speak – rapidly, with words blended together, varying emphasis and intonation, different accents, and often employing incorrect grammar.

- **Part-of-speech tagging**, also referred to as grammatical tagging, entails determining the grammatical classification of a specific word or piece of text based on its usage and context. Part-of-speech tagging identifies, for instance, "make" as a verb in "I can make a paper plane" and as a noun in "What make of car do you own?"

- **Word sense disambiguation** involves selecting the correct meaning of a word when it possesses multiple meanings, achieved through a process of semantic analysis to determine the most appropriate interpretation in the given context. For instance, word sense disambiguation helps distinguish the meaning of the verb "make" in "make the grade" (achieve) vs. "make a bet" (place).

- **Named entity recognition (NER)**, also known as NEM, identifies words or phrases as entities with specific roles. NER recognizes "Kentucky" as a location or "Fred" as a male name.

- **Coreference** resolution involves identifying whether two words refer to the same entity and is typically applied to determining the referent of a pronoun (e.g., "she" referring to "Mary"). It can also extend to identifying metaphors or idioms in the text, such as when "bear" refers not to the animal but to a large, hairy person.

- **Sentiment analysis** endeavors to extract subjective aspects such as attitudes, emotions, sarcasm, confusion, and suspicion from text.

- **Natural language generation**, often considered the counterpart of speech recognition or speech-to-text, revolves around the task of transforming structured information into human language.

Text Preprocessing and Feature Engineering

In the realm of analyzing and comprehending human language, natural language processing (NLP) employs an array of techniques and approaches. In the following, some foundational techniques commonly utilized in NLP are highlighted.

Tokenization

This process entails the segmentation of text into individual words, phrases, symbols, or other meaningful units known as tokens. Tokenization serves as a fundamental objective: to convey text in a format that retains its context while making it comprehensible to machines. Through the transformation of text into tokens, algorithms gain the capability to discern patterns effectively.

This capacity for pattern recognition is pivotal as it empowers machines to interpret and react to human inputs. For instance, when confronted with the word "running," a machine doesn't perceive it as a solitary entity; instead, it views it as a composition of tokens that it can dissect and extract meaning from.

Tokenization methods can vary depending on the level of granularity required and the specific demands of the task. These methods encompass a spectrum, ranging from disassembling text into individual words to breaking it down into characters or even smaller units:

- **Word Tokenization:** This method dissects text into discrete words, representing the most prevalent and effective approach. It works exceptionally well for languages like English, which have well-defined word boundaries.

- **Character Tokenization:** In character tokenization, the text is fragmented into individual characters. This approach proves valuable for languages lacking distinct word boundaries or for tasks demanding intricate analysis, such as spelling correction.

- **Subword Tokenization:** Striking a balance between word and character tokenization, subword tokenization divides text into units that are larger than a single character but smaller than a complete word. For example, "Chatbots" could be tokenized into "Chat" and "bots." This method finds particular utility in languages where meaning derives from combining smaller units or when handling words that are not found in the standard vocabulary in NLP tasks.

Parsing

In natural language processing (NLP), parsing refers to the process of analyzing the grammatical structure of a sentence or a piece of text to determine its syntactic and grammatical elements. The primary goal of parsing is to understand the hierarchical relationships between words and phrases within a sentence, identifying how they function within the sentence's structure.

Parsing involves the following key tasks:

- **Tokenization:** The text is first segmented into individual words or tokens.

- **Part-of-Speech Tagging:** Each token is tagged with its grammatical part of speech (e.g., noun, verb, adjective), which helps establish the role of each word in the sentence.

- **Dependency Parsing:** This step identifies the grammatical relationships between words in the sentence, establishing which words depend on or modify others. It creates a tree-like structure known as a dependency parse tree, where words are nodes connected by labeled edges representing grammatical relationships.

Parsing is essential in NLP for various applications, such as machine translation, information retrieval, sentiment analysis, and question answering. It enables computers to understand the syntactic structure of sentences, making it possible to extract meaningful information from text and generate coherent responses in natural language.

Lemmatization

This technique aims to reduce words to their base or root form, facilitating the grouping of various word forms with a common root. Lemmatization aims to find the base or dictionary form of a word, known as the "lemma," while preserving the word's actual meaning. It involves considering the word's part of speech (e.g., noun, verb, adjective) and context in order to perform a more accurate transformation.

Lemmatization results in valid words that can be found in a dictionary, making it suitable for tasks where the interpretability and meaningfulness of words are crucial. For example, the word "running" would be lemmatized to "run," and "better" would be lemmatized to "good."

Word Segmentation

Word segmentation involves the process of extracting individual words from a given text string. For instance, when a person scans a handwritten document into a computer, an algorithm can be employed to analyze the page and identify word boundaries, typically indicated by spaces between words.

Word Sense Disambiguation

Word sense disambiguation is a technique used to ascertain the intended meaning of a word within a given context. For instance, let's examine the sentence "The pig is in the pen." In this context, the word "pen" can have multiple meanings. An algorithm employing this approach can discern that in this instance, "pen" refers to a fenced-in enclosure for animals, as opposed to its alternative meaning, which denotes a writing instrument.

Sentence Boundary Detection

Sentence breaking, or sentence boundary detection, entails the task of identifying and marking sentence boundaries within lengthy textual passages. For example, when a natural language processing algorithm is presented with the text, "I played the game. I called Bill," it can discern the period as a marker that separates the two sentences, effectively breaking them apart.

Morphological Segmentation

Morphological segmentation dissects words into smaller units known as morphemes. For instance, the word "untestably" would be deconstructed into [[un[[test]able]]ly], with the algorithm recognizing "un," "test," "able," and "ly" as distinct morphemes. This technique proves particularly valuable in applications like machine translation and speech recognition.

Stemming

Stemming aims to isolate the root forms of words, especially those that contain inflections. For instance, in the sentence, "The boy played with the ball," the algorithm can identify that the root form of the word "played" is "play." This functionality proves useful when users seek to analyze a text for all occurrences of a specific word, including its various conjugations. The algorithm recognizes that these conjugated forms essentially represent the same word, despite variations in their lettering.

Employing this text-processing technique proves valuable in addressing issues of sparsity and standardizing vocabulary. It not only aids in minimizing redundancy, as inflected words and their word stems typically convey the same meaning, but it also enables NLP models to establish connections between inflected words and their corresponding word stems. This linkage enhances the model's comprehension of how these words are utilized in analogous contexts.

Stemming algorithms operate by identifying common prefixes and suffixes encountered in inflected words and truncating them from the word. Occasionally, this may yield word stems that do not correspond to actual words. Consequently, while this approach undoubtedly offers advantages, it is not exempt from its inherent limitations.

Named Entity Recognition (NER)

Named entity recognition (NER) stands as a pivotal subtask within the realm of natural language processing (NLP) and falls under the broader scope of information extraction. NER is dedicated to the classification of named entities into predefined categories, encompassing a wide array of entities such as personal names, organizations, geographic locations, medical codes, temporal expressions, quantities, monetary values, and more. In the field of NLP, comprehending these entities holds paramount importance, as they frequently encapsulate the most crucial pieces of information within a given text.

Named entity recognition (NER) operates as a vital intermediary that connects unstructured text with structured data, empowering machines to navigate extensive textual content and unearth valuable data fragments that are systematically categorized. By singling out specific entities from the vast expanse of words, NER revolutionizes the manner in which textual data is processed and leveraged.

The core mission of NER is to meticulously scan unstructured text, pinpoint distinct segments as named entities, and subsequently assign them to predefined categories. This transformation of unprocessed text into organized information enhances data's utility, facilitating various tasks such as data analysis, information retrieval, and the construction of knowledge graphs.

How Entities Are Recognized

The inner workings of named entity recognition (NER) can be deconstructed into several sequential stages:

- **Tokenization:** Before delving into entity identification, the text undergoes tokenization, which involves segmenting it into tokens, which can take the form of words, phrases, or even entire sentences.

- **Entity Identification:** The next step entails the identification of potential named entities within the text. This process relies on various linguistic rules or statistical methods to detect entities. It encompasses recognizing patterns such as capitalization in names (e.g., "Bill Gates") or adherence to specific formats, such as dates.

- **Entity Classification:** Once entities are pinpointed, they are systematically categorized into predefined classes such as "Person," "Organization," or "Location." This classification is often achieved through machine learning models that have been trained on annotated datasets. For instance, "Bill Gates" would be classified as a "Person," while "Microsoft" would be designated as an "Organization."

- **Contextual Analysis:** NER systems frequently take into account the contextual surroundings to enhance accuracy. For example, in the sentence "Apple released a new iPhone," the contextual cues assist the system in recognizing "Apple" as an organization rather than a fruit.

- **Postprocessing:** Following the initial recognition and classification phases, postprocessing techniques may be employed to refine the results. This can encompass the resolution of ambiguities, consolidation of multi-token entities, or the utilization of knowledge bases to augment entity information.

NLP and Feature Engineering

Feature extraction in NLP is a critical process where raw textual content is converted into numerical representations that are more accessible for machine learning models. There is a range of methods for feature extraction in NLP, each offering unique advantages and limitations. For data scientists, it is imperative to be well versed in these various techniques and understand when to apply them effectively.

This step is vital in NLP as feature engineering plays an essential role in natural language processing (NLP) applications. It involves defining features that characterize an NLP task, shaping the way a machine learning algorithm understands and processes text data, and it greatly influences the performance of the models designed for specific tasks.

Python and the Natural Language Toolkit (NLTK)

The Python programming language offers a diverse array of tools and libraries tailored for addressing specific natural language processing (NLP) tasks. Within this landscape, you'll find the Natural Language Toolkit, commonly known as NLTK, which stands as an open source assemblage of libraries, utilities, and educational resources designed to facilitate the construction of NLP applications.

The NLTK encompasses libraries catering to various NLP tasks mentioned previously, and it also includes specialized libraries for subtasks such as sentence parsing, word segmentation, stemming, lemmatization (techniques for reducing words to their root forms), and tokenization (breaking down phrases, sentences, paragraphs, and passages into tokens that enhance the computer's comprehension of the text). Additionally, the NLTK houses libraries that enable the implementation of advanced capabilities like semantic reasoning, which involves deriving logical conclusions from facts extracted from textual sources.

In the next paragraphs, some examples of feature engineering are provided.

Install the latest version of NLTK by typing `pip install nltk`. This is valid for all examples with NLTK as shown here.

Tokenization

```
import nltk
from nltk.tokenize import word_tokenize
nltk.download('punkt')

text = "The quick brown fox jumped over the lazy dog."
tokens = word_tokenize(text)
print(tokens)

Output:
['The', 'quick', 'brown', 'fox', 'jumped', 'over', 'the', 'lazy', 'dog', '.']
```

Stop Word Removal

```
import nltk
from nltk.corpus import stopwords
from nltk.tokenize import word_tokenize
```

```
nltk.download('stopwords')

text = "The quick brown fox jumped over the lazy dog."
stop_words = set(stopwords.words('english'))
tokens = word_tokenize(text)
filtered_tokens = [token for token in tokens if not token in stop_words]
print(filtered_tokens)

Output:
['The', 'quick', 'brown', 'fox', 'jumped', 'lazy', 'dog', '.']
```

Stemming

```
import nltk
from nltk.stem import PorterStemmer
from nltk.tokenize import word_tokenize
nltk.download('punkt')

text = "The quick brown foxes jumped over the lazy dogs."
stemmer = PorterStemmer()
tokens = word_tokenize(text)
stemmed_tokens = [stemmer.stem(token) for token in tokens]
print(stemmed_tokens)
```

Output

```
['the', 'quick', 'brown', 'fox', 'jump', 'over', 'the', 'lazi', 'dog', '.']
```

Lemmatization

```
import nltk
from nltk.stem import WordNetLemmatizer
from nltk.tokenize import word_tokenize
nltk.download('wordnet')

text = "The quick brown foxes jumped over the lazy dogs."
lemmatizer = WordNetLemmatizer()
tokens = word_tokenize(text)
lemmatized_tokens = [lemmatizer.lemmatize(token) for token in tokens]
print(lemmatized_tokens)
```

Output:

['The', 'quick', 'brown', 'fox', 'jumped', 'over', 'the', 'lazy', 'dog', '.']

N-grams for NLP Feature Engineering

```
import nltk
from nltk.util import ngrams

text = "The quick brown foxes jumped over the lazy dogs."
tokens = nltk.word_tokenize(text)
bigrams = ngrams(tokens, 2) trigrams = ngrams(tokens, 3)
print(list(bigrams))
print(list(trigrams))
```

Output:

[('The', 'quick'), ('quick', 'brown'), ('brown', 'foxes'), ('foxes', 'jumped'), ('jumped', 'over'), ('over', 'the'), ('the', 'lazy'), ('lazy', 'dogs'), ('dogs', '.')] [('The', 'quick', 'brown'), ('quick', 'brown', 'foxes'), ('brown', 'foxes', 'jumped'), ('foxes', 'jumped', 'over'), ('jumped', 'over', 'the'), ('over', 'the', 'lazy'), ('the', 'lazy', 'dogs'), ('lazy', 'dogs', '.')]

Part-of-Speech (POS) Tagging

```
import nltk
nltk.download('averaged_perceptron_tagger')
text = "The quick brown foxes jumped over the lazy dogs."
tokens = nltk.word_tokenize(text)
pos_tags = nltk.pos_tag(tokens)
print(pos_tags)
```

Output:

[('The', 'DT'), ('quick', 'JJ'), ('brown', 'NN'), ('foxes', 'NNS'), ('jumped', 'VBD'), ('over', 'IN'), ('the', 'DT'), ('lazy', 'JJ'), ('dogs', 'NNS'), ('.', '.')]

Named Entity Recognition (NER)

```
import nltk
nltk.download('words')
nltk.download('maxent_ne_chunker')
nltk.download('averaged_perceptron_tagger')

text = "John Smith works at Google in New York City."
tokens = nltk.word_tokenize(text)
pos_tags = nltk.pos_tag(tokens)
ner_tags = nltk.ne_chunk(pos_tags)
print(ner_tags)
```

Output:

(S (PERSON John/NNP) (PERSON Smith/NNP) works/VBZ at/IN (ORGANIZATION Google/NNP) in/IN (GPE New/NNP York/NNP City/NNP) ./.)

TF-IDF

```
from sklearn.feature_extraction.text import TfidfVectorizer

corpus = [ "The quick brown fox jumps over the lazy dog.",
          "The quick brown foxes jump over the lazy dogs and cats.",
          "The lazy dogs and cats watch the quick brown foxes jump over
          the moon."]

tfidf_vectorizer = TfidfVectorizer()
tfidf_matrix = tfidf_vectorizer.fit_transform(corpus)
print(tfidf_matrix.toarray())
```

Output:

[[0. 0. 0. 0.51785612 0. 0. 0. 0. 0. 0. 0.68091856 0.51785612 0.]
 [0. 0. 0. 0.46519584 0. 0. 0.59817854 0. 0. 0. 0. 0.46519584 0.59817854]
 [0.33682422 0.33682422 0.33682422 0.30794004 0.33682422 0.33682422 0.
 0.33682422 0.33682422 0.33682422 0. 0.30794004 0.]]

Table 1-1 lists some common feature extraction techniques for natural language processing based on NLTK.

Table 1-1. *Common feature extraction techniques for NLP*

Technique	Main Features	Use Cases	Size and Complexity
CountVectorizer	Converts text to matrix of word counts	Text classification, topic modeling	Simple and fast, suitable for small to medium-sized datasets
TF-IDF	Assigns weights to words based on importance	Information retrieval, text classification	More complex and computationally expensive, suitable for medium to large-sized datasets
Word embeddings	Vector representation of words based on semantics and syntax	Text classification, information retrieval	Can handle large datasets, computationally expensive to train
Bag of words	Represents text as a vector of word frequencies	Text classification, sentiment analysis	Simple and fast, suitable for small to medium-sized datasets
Bag of n-grams	Captures frequency of sequences of n words	Text classification, sentiment analysis	Size and complexity depend on the size of the n-grams and the dataset
Hashing Vectorizer	Maps words to fixed-size feature space using hashing function	Large-scale text classification, online learning	Suitable for large datasets, memory efficient, may suffer from hash collisions
Latent Dirichlet Allocation (LDA)	Identifies topics in corpus and assigns probability distribution to each document	Topic modeling, content analysis	Suitable for medium to large-sized datasets, computationally expensive
Non-negative matrix factorization (NMF)	Decomposes document-term matrix into lower-dimensional parts	Topic modeling, content analysis	Suitable for medium-sized datasets, computationally expensive

(*continued*)

Table 1-1. (*continued*)

Technique	Main Features	Use Cases	Size and Complexity
Principal component analysis (PCA)	Reduces dimensionality of document-term matrix	Text visualization, text compression	Suitable for large datasets, computationally expensive
Part-of-speech (POS) tagging	Assigns part of speech tag to each word in text	Named entity recognition, text classification	Requires additional processing, suitable for small to medium-sized datasets

Word Embeddings and Semantic Understanding

Word embedding is a key concept in natural language processing (NLP), where words are transformed into real-valued vectors for textual analysis. This advancement in NLP has significantly enhanced computer capabilities in comprehending text. It's a major leap in deep learning, addressing complex NLP challenges effectively.

A major challenge for natural language processing (NLP) is its inability to interpret words with multiple meanings based on context. Contextual semantic analysis plays a pivotal role in clarifying such ambiguities, enhancing the precision of text-based NLP applications. Let's explore the significance of disambiguation in NLP.

This section takes a closer look at both concepts.

Word Embedding

In word embedding, both words and documents are converted into numerical vectors. This conversion ensures that words with similar meanings share similar vector representations. Its technique captures meanings in a compressed dimensional space, offering a faster alternative to traditional, more complex graph embedding models like WordNet.

The vectors are then utilized by machine learning models, maintaining the text's semantic and syntactic integrity. The transformed data is processed by NLP algorithms, which efficiently interpret these representations.

There are generally three types of vectors under this category:

- Count Vector
- TF-IDF Vector
- Co-occurrence Vector

Word embeddings are capable of training advanced deep learning models such as GRU, LSTM, and Transformers. These models have shown remarkable success in various NLP tasks, including sentiment analysis, named entity recognition, and speech recognition.

The effectiveness of word embedding has boosted the popularity of ML in NLP, making it a highly favored field among developers. For example, a 50-value word embedding can represent 50 distinct characteristics. Many opt for preexisting word embedding models such as Flair, fastText, and spaCy, among others.

Common techniques for word embeddings, some of them already featured in this book, are as follows:

- Term frequency-inverse document frequency (TF-IDF)
- Bag of words (BOW)
- Word2Vec – Capturing Semantic Information
- GloVe – Global Vectors for Word Representation
- BERT – Bidirectional Encoder Representations from Transformers

Benefits

1. Word embeddings excel in grasping the meaning and context of words, leading to enhanced accuracy in text analysis and predictions.

2. They offer a more effective and scalable method to represent words compared to traditional bag-of-words models.

3. Word embeddings can be honed using extensive text data, enabling them to discern intricate linguistic nuances and connections.

Limitations

1. The effectiveness of word embeddings might be limited for certain text types or languages.

2. They might not fully capture all subtleties of meaning and context, as they rely on statistical patterns in data and may not accurately represent real semantic links between words.

3. Word embeddings can demand significant computational resources and memory, particularly with larger data sets.

Example:

Before running the example, run the command `pip install gensim` to get the latest version of Gensim.

```
from gensim.models import Word2Vec

sentences = [ "The quick brown fox jumps over the lazy dog".split(),
              "The lazy dog watches the quick brown fox".split(),
              "The quick brown cat jumps over the lazy dog".split(),
              "The lazy dog watches the quick brown cat".split() ]

model = Word2Vec(sentences, min_count=1)
print(model.wv['quick'])
```

Output:

The expected output is too big; that's why just an exception is presented:

```
array([[ 0.03135875,  0.03640932, -0.00031054,  0.04873694, -0.03376802],
       [ 0.00243857, -0.02919209, -0.01841091, -0.03684188,  0.02765827],
       [-0.01245669, -0.01057661, -0.04422194, -0.0317696 , -0.00031216]],
      dtype=float32)
```

Semantic Understanding

Machines often struggle to grasp the inherent meaning in words, sentences, and documents. Techniques like word sense disambiguation and recognition of meanings can improve machines' understanding of language.

The components of semantic analysis include the following:

- Lexical analysis, which involves scanning character streams, identifying lexemes, and converting them into machine-readable tokens.

- Grammatical analysis that links lexemes in sequence, applying formal grammar for part-of-speech tagging.

- Syntactical analysis, which examines syntax and enforces grammar rules, aiding in contextual understanding at both the word and sentence levels.

- Semantic analysis combines these elements to discern word meanings and interpret sentence structures, enabling machine comprehension similar to human understanding.

The Importance of Semantic Analysis in NLP

Extracting business intelligence from language data is crucial yet challenging for many organizations, especially in analyzing unstructured data. This obstacle often hampers AI projects focused on language-intensive tasks.

Communications, whether tweets, LinkedIn posts, or website comments, may hold critical information. Successfully capturing and interpreting this data is vital for competitive advantage. However, capturing data is simpler than understanding its contents, especially on a large scale.

In NLP, basic techniques can identify words, but semantics provides contextual meaning (like distinguishing between "apple" as a fruit or a company).

Human language comprehension is almost automatic, leveraging our neural networks. We interpret texts based on our knowledge of language and the text's concepts. Machines, however, cannot mimic this process naturally.

Some technologies give the illusion of understanding text, often relying on keyword matching or frequency-based approaches, typical in computational linguistics or statistical NLP. Such methods have limitations as they don't truly seek to comprehend meaning.

In contrast, semantic analysis is key to achieving high accuracy in text analysis. For instance, in text summarization, which condenses large texts into concise summaries, the effectiveness depends on the machine's language comprehension.

Semantic Analysis Within a Semantic System

A semantic system integrates entities, concepts, relations, and predicates, enriching language context for more accurate machine interpretation. Semantic analysis extracts meaning from language, foundational for semantic systems in enhancing machine interpretation.

Key elements of semantic analysis supporting language understanding include the following:

- **Hyponymy:** A general term
- **Homonymy:** Lexical terms with identical spelling but different meanings
- **Polysemy:** Terms with identical spelling and related meanings
- **Synonymy:** Different-spelled lexical terms with similar meanings
- **Antonymy:** Lexical terms with opposite meanings
- **Meronomy:** The relationship between a term and a broader entity

Grasping these concepts is vital for NLP applications aimed at extracting insights and information and supporting automated processing and conversational systems like chatbots.

Sentiment Analysis and Text Classification with Python

Sentiment analysis, also known as opinion mining, is a natural language processing (NLP) technique used to determine the sentiment, emotional tone, or subjective opinions expressed in a piece of text. A key application within NLP is text classification, which involves labeling or categorizing text. Let's review each technique.

Sentiment Analysis

The primary goal of sentiment analysis is to automatically assess the sentiment or emotional state conveyed by the text, whether it is positive, negative, neutral, or a more nuanced sentiment like joy, anger, sadness, or surprise.

Key aspects of sentiment analysis in NLP include the following:

- **Sentiment Classification:** Sentiment analysis typically involves classifying text into one or more sentiment categories. The most common categories are as follows:

  - **Positive:** Indicates a favorable sentiment, such as satisfaction, happiness, or approval.

  - **Negative:** Signifies an unfavorable sentiment, including dissatisfaction, disappointment, or criticism.

  - **Neutral:** Represents a lack of strong sentiment or a balanced viewpoint.

  - **Mixed or Neutral:** Some sentiment analysis systems also provide mixed or nuanced sentiment categories to capture more complex emotions.

- **Text Processing:** Sentiment analysis algorithms preprocess text by tokenizing it (splitting it into words or phrases), removing stop words, and applying techniques like stemming or lemmatization to normalize the text.

- **Feature Extraction:** Sentiment analysis systems extract features from the text, which may include words, phrases, or even sentence structure, that are indicative of sentiment.

Each of these techniques plays a pivotal role in enabling computers to process and make sense of human language, serving as the fundamental building blocks for more advanced NLP applications.

Example:

Before running the example, run the command `pip install textblob==0.18.0`.

```
from textblob import TextBlob

def analyze_sentiment(text):
    testimonial = TextBlob(text)
    polarity = testimonial.sentiment.polarity
    subjectivity = testimonial.sentiment.subjectivity
```

```
    if polarity > 0:
        return "Positive Sentiment"
    elif polarity == 0:
        return "Neutral Sentiment"
    else:
        return "Negative Sentiment"
# Example usage
text = "I love this phone. It has an amazing camera!"
result = analyze_sentiment(text)
print(f"Sentiment Analysis Result: {result}")
```

Text Classification

In today's increasingly digital world, understanding and processing vast volumes of data is crucial. Natural language processing (NLP), a dynamic and influential domain in computer science, addresses this need.

For instance, emails can be classified as spam or not, tweets as positive or negative, and articles as relevant or irrelevant to a specific subject. This guide aims to introduce the fundamentals of text classification, paving the way for you to develop your own models.

Advantages of Text Classification

Text classification has diverse applications and is instrumental in several areas:

- **Document Categorization:** Automates the process of sorting documents, aiding in efficient information retrieval

- **Outcome Prediction:** Predicts future events, like forecasting stock market responses from news articles

- **Pattern Discovery:** Uncovers hidden trends, such as identifying customer preferences in marketing campaigns

Varieties of Text Classification

There are mainly two types: supervised and unsupervised. Supervised classification involves models trained on pre-labeled data, while unsupervised classification does not use labels but learns from the data itself, identifying categories based on similarities. Supervised classification typically yields higher accuracy but requires more resources. Unsupervised classification, though less precise, is useful when labeled data is unavailable.

Mechanics of Text Classification

Text classification involves categorizing text based on its content. This process usually starts by defining a set of categories or classes and then identifying keywords or phrases indicative of these classes. Subsequently, texts are assigned to these classes, either manually or automatically. The final step involves analyzing the classified data, which can range from simple counts to complex machine learning analyses.

Challenges in Text Classification

Text classification faces challenges due to the unstructured and noisy nature of text data. High dimensionality of textual data and the requirement for deep language comprehension by machines add to the complexity. Advanced NLP techniques like sentiment analysis or topic modeling are often required for deeper understanding.

Applications of Text Classification

Text classification serves many purposes, such as detecting spam, analyzing sentiment, categorizing news, and identifying topics. As a supervised learning task, it necessitates a labeled dataset for training models to classify text accurately. This makes it an integral tool in various domains, streamlining processes and offering insights into large volumes of text data.

Example of News Text Classification:

```
from sklearn.feature_extraction.text import TfidfVectorizer
from sklearn.linear_model import LogisticRegression

# Load the dataset
news = pd.read_csv('news.csv')
```

```
# Split the data into training and testing sets
X_train, X_test, y_train, y_test = train_test_split(news['text'],
news['category'], test_size=0.2)

# Create a TfidfVectorizer to convert text to numerical features
vectorizer = TfidfVectorizer()
X_train = vectorizer.fit_transform(X_train)
X_test = vectorizer.transform(X_test)

# Train a logistic regression model
model = LogisticRegression()
model.fit(X_train, y_train)

# Evaluate the model on the test set
score = model.score(X_test, y_test)
print('Accuracy:', score)
```

Example of News Text Classification:
Before executing this example, run the following commands:

- **pip install pandas**
- **pip install scikit-learn**

```
from sklearn.datasets import fetch_20newsgroups
from sklearn.feature_extraction.text import TfidfVectorizer
from sklearn.linear_model import LogisticRegression
from sklearn.model_selection import train_test_split
import pandas as pd

# Load the dataset
news = fetch_20newsgroups(subset='all', categories=None, shuffle=True,
random_state=42)

# Split the data into training and testing sets
X_train, X_test, y_train, y_test = train_test_split(news.data, news.target,
test_size=0.2, random_state=42)

# Create a TfidfVectorizer to convert text to numerical features
vectorizer = TfidfVectorizer()
```

```
X_train = vectorizer.fit_transform(X_train)
X_test = vectorizer.transform(X_test)

# Train a logistic regression model
model = LogisticRegression(max_iter=1000)
model.fit(X_train, y_train)

# Evaluate the model on the test set
score = model.score(X_test, y_test)
print('Accuracy:', score)
```

Exploring Approaches in Natural Language Processing (NLP)

Let's delve into some common methodologies employed in the field of natural language processing.

Supervised NLP

Supervised NLP methods involve training the software using a dataset where both inputs and corresponding outputs are labeled or known. Initially, the program processes extensive sets of established data, learning to generate correct outputs for any unfamiliar input. For instance, companies utilize supervised NLP to train tools for categorizing documents according to predefined labels.

Unsupervised NLP

Unsupervised NLP relies on statistical language models to anticipate patterns when provided with input that lacks labeling. An example of this is the autocomplete feature in text messaging, which suggests appropriate words for a sentence by observing the user's input.

Natural Language Understanding

Natural language understanding (NLU) represents a subset of NLP that concentrates on deciphering the meaning concealed within sentences. NLU equips software to discern analogous meanings in varying sentences or to comprehend words with multiple interpretations.

Natural Language Generation

Natural language generation (NLG) is concerned with crafting conversational text in a manner akin to human communication, guided by specific keywords or topics. For instance, an intelligent chatbot equipped with NLG capabilities can engage with customers in a manner akin to human customer support personnel.

Statistical NLP, Machine Learning, and Deep Learning

The earliest iterations of NLP applications were manually crafted, rule-based systems capable of executing specific NLP functions. However, they struggled to adapt to the ever-growing number of exceptions and the escalating volume of textual and spoken data.

This is where statistical NLP made its entrance, blending computer algorithms with machine learning and deep learning models. It enables the automatic extraction, classification, and labeling of components within textual and voice data, assigning statistical probabilities to potential meanings for these components.

In contemporary times, deep learning models and learning techniques rooted in convolutional neural networks (CNNs) and recurrent neural networks (RNNs) have empowered NLP systems to "learn" progressively as they operate, extracting increasingly precise meanings from vast quantities of unprocessed, unstructured, and unlabeled text and voice datasets.

NLP Challenges

Despite its progress, natural language processing (NLP) grapples with several formidable challenges arising from the intricate and nuanced nature of human language. Here are some of these prominent NLP challenges:

- **Ambiguity:** Human language frequently exhibits ambiguity, where words carry multiple meanings. This complexity poses a substantial hurdle for NLP models when attempting to discern the precise interpretation of words within diverse contexts.

- **Context:** Accurate comprehension hinges on grasping the context in which words are employed. NLP faces an ongoing struggle in effectively deciphering the contextual nuances that underlie language usage.

- **Precision:** Traditional computing necessitates humans to communicate in a programming language characterized by exactness, clarity, and structured syntax. In stark contrast, human speech often lacks precision, frequently featuring ambiguity, influenced by factors such as slang, regional dialects, and social context.

- **Tone and Inflection:** NLP remains a work in progress concerning tone, inflection, and semantic analysis. Detecting elements like sarcasm proves challenging, as it necessitates an understanding of word choice and context. Additionally, subtle yet significant tone variations in spoken language may elude NLP algorithms. The influence of word stress or syllable emphasis on sentence meaning, as well as variations in speech tone due to different accents, presents further complexities for algorithmic parsing.

- **Evolving Language Usage:** The dynamic nature of language, marked by continual evolution in its usage patterns, presents an ongoing challenge for NLP. Although language follows established rules, these are not immutable, subject to shifts over time. Computational rules that currently hold relevance may become outdated as real-world language evolves.

Delineating NLP Basics from LLM Capabilities

It is crucial to distinguish between the foundational principles of NLP and the remarkable capabilities of LLMs. Traditional NLP encompasses a diverse spectrum of techniques, spanning from rule-based systems that rely on manually crafted rules to statistical methods that leverage probabilistic models and machine learning algorithms to learn patterns from data. These traditional NLP techniques are designed and optimized to address specific NLP tasks, such as named entity recognition, machine translation, or sentiment analysis.

In contrast, LLMs represent a subset of NLP models that are characterized by their deep learning architecture, which typically involves layers of artificial neural networks. This deep learning approach enables LLMs to learn from extensive data corpora and generalize their knowledge across a wide range of tasks. Unlike traditional NLP techniques, LLMs do not require explicit programming or feature engineering for each specific task. Instead, they can learn intricate relationships and patterns directly from the data, making them exceptionally versatile.

However, this versatility comes at a cost. LLMs are computationally intensive, requiring substantial resources for training and deployment. Additionally, they are data-hungry, relying on vast amounts of text data to achieve optimal performance. These factors pose challenges in terms of computational infrastructure and data availability, which must be carefully considered when employing LLMs in real-world applications.

Understanding the distinction between traditional NLP techniques and LLMs is vital for selecting the most appropriate approach for a given task. Traditional NLP techniques may be more suitable when the task is well defined and the required data is limited. On the other hand, LLMs offer unparalleled versatility and the potential to tackle a broader range of tasks, albeit with higher computational demands and data requirements.

Contrasting Traditional NLP Techniques with LLMs

The fundamental distinction between conventional NLP techniques and LLMs lies in their respective approaches and the scope of their capabilities. Traditional NLP methods are typically designed to excel at specific tasks, such as machine translation, sentiment analysis, or named entity recognition. These methods rely on carefully crafted features and algorithms that are tailored to each particular task. While this approach can be highly effective for well-defined tasks, it often requires substantial manual effort to engineer the necessary features and algorithms.

In contrast, LLMs take a much broader, data-driven approach to NLP. Instead of being explicitly programmed for specific tasks, LLMs are trained on vast amounts of text data and learn to identify patterns and relationships within the language. This allows them to perform a wide range of tasks, even those for which they were not explicitly trained. For example, an LLM can be used for text generation, question answering, summarization, and even creative writing.

However, this versatility comes at a price. LLMs require significantly more computational resources and data than traditional NLP methods. Training an LLM typically involves using large-scale datasets and powerful hardware, such as graphical processing units (GPUs). Additionally, LLMs can be challenging to control and can sometimes generate nonsensical or biased output.

Table 1-2 summarizes the key differences between conventional NLP techniques and LLMs.

Table 1-2. Conventional NLP techniques vs. LLMs

Feature	Conventional NLP Techniques	LLMs
Approach	Task-specific, engineered features and algorithms	Data-driven, learns patterns from text
Scope	Limited to specific tasks	Wide range of tasks
Computational resources	Less computationally expensive	More computationally expensive
Data requirements	Smaller datasets	Larger datasets
Controllability	More controllable	Less controllable

Despite these challenges, LLMs represent a major breakthrough in the field of NLP and are rapidly advancing the state of the art in many applications. As these models continue to develop and improve, it is expected to see them play an increasingly important role in our lives.

While the bedrock of NLP provides the core techniques and foundational knowledge necessary for tackling language processing tasks, LLMs herald a groundbreaking leap forward. LLMs offer an unrivaled breadth of versatility and exceptional performance across a diverse spectrum of tasks. However, it is important to acknowledge that LLMs come with their own set of unique requirements and challenges.

For those aspiring to embark on a journey into the captivating realm of language processing and artificial intelligence, a comprehensive understanding of both NLP principles and the capabilities of LLMs is of paramount importance. By mastering the fundamentals of NLP, one gains insights into the essential techniques, algorithms, and mathematical underpinnings that facilitate the processing, understanding, and generation of human language.

On the other hand, exploring the world of LLMs unveils their remarkable ability to learn from vast text corpora, enabling them to perform a multitude of language-related tasks, ranging from text generation and translation to question answering and summarization. LLMs have demonstrated impressive results in various domains, such as healthcare, finance, and customer service, showcasing their potential to revolutionize industries.

However, it is crucial to recognize that LLMs are not without limitations. They may exhibit biases inherited from the training data, struggle with common-sense reasoning, and occasionally generate inaccurate or nonsensical responses. Understanding these challenges is essential for responsible and effective deployment of LLMs in real-world applications.

By cultivating a deep understanding of both NLP fundamentals and LLM capabilities, aspiring practitioners and researchers will be well equipped to navigate the complexities of language processing, contribute to the ongoing advancements in the field, and unlock the full potential of LLMs in driving innovation and solving real-world problems.

Summary

This chapter provides a comprehensive overview of the development and impact of large language models (LLMs) within the field of artificial intelligence. The chapter traces the progression from early statistical and rule-based models to contemporary transformer-based models with billions of parameters. It discusses the emergence and achievements of notable LLMs such as OpenAI's ChatGPT, Google Bard, and Meta's LLaMA, highlighting their contributions to natural language processing (NLP).

The chapter concludes by discussing the era of multimodal learning, where LLMs are integrated with diverse data types such as text, images, and audio.

In the following chapter, we will explore in detail the following:

- How large language models work
- Benefits and challenges of LLMs in various domains
- Why are LLMs so popular?

CHAPTER 2

What Are Large Language Models?

The extraordinary capacity of human beings to communicate through language begins to develop in early childhood and continues to evolve throughout their lives. However, machines lack an inherent ability to comprehend and communicate in human language unless equipped with powerful AI algorithms.[1] The long-standing research challenge and aspiration have been to enable machines to attain human-like reading, writing, and communication skills.

The emergence of large language models (LLMs) can be attributed to advances in deep learning (DL) techniques, the availability of substantial computational resources, and the abundance of training data. These models, often pre-trained on extensive web corpora, possess the capability to grasp intricate patterns, linguistic subtleties, and semantic relationships. Fine-tuning these models for specific tasks has yielded promising results, achieving state-of-the-art performance across various benchmarks.

LM's Development Stages

Enhancing the language intelligence of machines is a critical objective achieved through language modeling (LM). Broadly, LM entails modeling the probability of word sequences to predict future likelihoods. The LM research has garnered widespread attention and has progressed through four notable developmental stages.

---

[1] W. X. Zhao, K. Zhou, J. Li, T. Tang, X. Wang, Y. Hou, Y. Min, B. Zhang, J. Zhang, Z. Dong, et al., "A survey of large language models," arXiv preprint arXiv:2303.18223, 2023

CHAPTER 2   WHAT ARE LARGE LANGUAGE MODELS?

The initial milestone in LM was the advent of **Statistical Language Models (SLMs)**, including n-gram models.[2] These models gauge the probability of the next word in a sequence based on the frequency of previous n-grams of words.[3] For instance, a bigram model leverages the frequency of word pairs to estimate the probability of the succeeding word.

The second phase of LM development introduced **Neural Language Models (NLMs)**, also known as neural language modeling. This approach employs neural networks to predict the probability distribution of the next word given the preceding words in the sequence. Recurrent neural networks (RNNs) and their variations like Long Short-Term Memory (LSTM) and Gated Recurrent Units (GRUs) are commonly utilized in this paradigm.[4]

The third stage of LM evolution encompasses the emergence of contextualized word embeddings, termed **Pre-trained Language Models (PLMs)**. These models employ neural networks to acquire a vector representation of words that considers the context in which the word appears. Examples of contextualized word embeddings include ELMo[5] and BERT.

The fourth phase of language model (LM) advancement marked the inception of extensive pre-training language models known as **Large Language Models (LLMs).**[6] These models, exemplified by GPT-3 and GPT-4, possess the capability to excel in various natural language processing (NLP) tasks. They undergo training on vast volumes of text data and can be fine-tuned for specific tasks like language translation or question answering.

To sum up, these four developmental stages in LM (visualized in Figure 2-1) signify substantial progress in the field, with each stage building upon its predecessor and pushing the boundaries of what machines can accomplish in both NLP and computer vision.

---

[2] J. Gao and C.-Y. Lin, "Introduction to the special issue on statistical language modeling," 2004

[3] A. Stolcke, "Srilm-an extensible language modeling toolkit," in Seventh international conference on spoken language processing, 2002

[4] S. Kombrink, T. Mikolov, M. Karafiát, L. Burget, "Recurrent neural network based language modeling in meeting recognition," in Interspeech, vol. 11, pp. 2877–2880, 2011

[5] M. Peters, M. Neumann, M. Iyyer, M. Gardner, C. Clark, K. Lee, and L. Zettlemoyer, "Deep contextualized word representations. arxiv 2018," arXiv preprint arXiv:1802.05365, vol. 12, 2018

[6] M. Shanahan, "Talking about large language models," arXiv preprint arXiv:2212.03551, 2022

CHAPTER 2   WHAT ARE LARGE LANGUAGE MODELS?

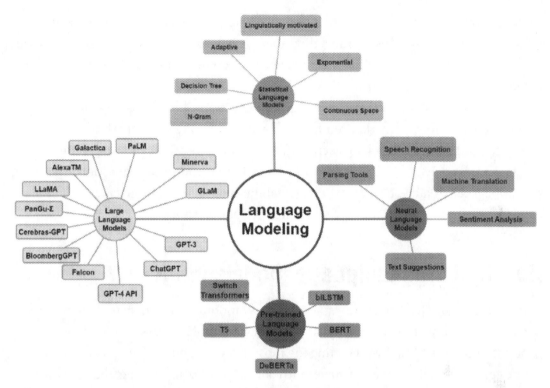

Figure 2-1. Different types of language modeling (Source: www.researchgate. net/figure/Types-of-language-modeling_fig1_372258530. License: Creative Commons Attribution 4.0 International)

A large language model stands as an advanced iteration of language models, undergoing training through deep learning methodologies on extensive text datasets. These models demonstrate the capability to generate text that closely resembles human expression and excel in diverse natural language processing tasks.

In contrast, the definition of a language model revolves around the idea of assigning probabilities to word sequences through the analysis of text corpora. The complexity of a language model can vary, ranging from basic n-gram models to more intricate neural network models. However, the term "large language model" typically pertains to models employing deep learning techniques and featuring a substantial number of parameters, ranging from millions to billions. Such models excel in capturing intricate language patterns, producing text that often mirrors human-authored content.

A "large language model," often embodied in a massive transformer model, is typically too substantial to operate on a single computer. Consequently, it is offered as a service through an API or web interface. These models undergo training on extensive text data from diverse sources, such as books, articles, websites, and various written content forms. Through this training, they analyze statistical relationships among words, phrases, and sentences, enabling them to generate coherent and contextually relevant responses to prompts or queries.

For example, GPT-3, the large language model within ChatGPT, is an example trained on vast Internet text data, endowing it with the ability to comprehend multiple languages and possess knowledge across diverse topics. Its proficiency in producing text across various styles, including translation, text summarization, and question answering, may appear remarkable. However, these capabilities operate using specific "grammars" aligned with prompts, explaining their impressive performance.

How Do Large Language Models Work?

Large language models operate by analyzing vast amounts of text data to learn patterns and relationships within language. Using advanced neural network architectures, they generate human-like text by predicting the most likely sequence of words based on the context provided. This process involves intricate layers of computation and massive datasets to achieve their impressive capabilities.

Large language models like GPT-3, utilizing a transformer architecture, function through the following simplified process:

- **Learning from Abundant Text:** These models initiate by assimilating a vast amount of text from the Internet, akin to learning from an extensive library of information.

- **Innovative Architecture:** Employing a distinctive structure called a transformer, they gain the ability to understand and retain substantial amounts of information.

- **Word Breakdown:** The models dissect sentences into smaller components, effectively breaking down words. This segmentation enhances their efficiency in working with language.

- **Understanding Sentence Structure:** Distinguishing themselves from simple programs, these models comprehend individual words and their relationships within a sentence, grasping the entire context.

- **Specialized Training:** After the initial learning phase, models can undergo further training on specific topics, enhancing their proficiency in tasks such as answering questions or writing about particular subjects.

- **Task Execution:** When presented with a prompt, these models utilize their acquired knowledge to generate responses, resembling an intelligent assistant capable of comprehending and producing text.

Overall Architecture of Large Language Models

The foundational structure of large language models is primarily composed of various neural network layers, including recurrent layers, feedforward layers, embedding layers, and attention layers.

These layers collaborate to process input text and formulate output predictions:

1. **The embedding layer** serves to convert each word in the input text into a high-dimensional vector representation. These embeddings encapsulate both semantic and syntactic information about the words, aiding the model in grasping the contextual nuances.

2. **The feedforward layers in large language models** encompass multiple fully connected layers that apply nonlinear transformations to the input embeddings. These layers contribute to the model's capacity to discern higher-level abstractions within the input text.

3. Designed to interpret information sequentially from the input text, **the recurrent layers** of LLMs maintain a hidden state updated at each time step. This dynamic allows the model to capture dependencies between words in a sentence effectively.

4. An integral component of LLMs, the **attention mechanism** empowers the model to selectively focus on distinct portions of the input text. This mechanism enhances the model's ability to attend to the most relevant segments of the input text, resulting in more precise predictions.

CHAPTER 2   WHAT ARE LARGE LANGUAGE MODELS?

In-Depth Architecture of the LLMs

At their core, large language models (LLMs) represent a form of artificial intelligence designed to emulate human intelligence. They operate by employing sophisticated statistical models and deep learning techniques to analyze extensive volumes of text data. These models learn intricate patterns and relationships inherent in the data, enabling them to produce new content that closely mimics the style and characteristics of a particular author or genre.

The process commences with pre-training, where the LLM is exposed to a vast corpus of text from diverse sources like books, articles, and websites. Through unsupervised learning, the model predicts the next word in a sentence based on the context of preceding words, fostering an understanding of grammar, syntax, and semantic relationships.

The LLM's pre-training pipeline involves the first step of collecting corpus sources, categorized into general and specialized data. Subsequently, data preprocessing becomes crucial to generate the pre-training corpus, involving the removal of noisy, redundant, unnecessary, and potentially harmful material.

The second stage encompasses quality filtering to eliminate low-quality and undesired data using techniques like language filtering, statistical filtering, and keyword filtering. The third stage involves deduplication, addressing the finding that duplicate data reduces diversity in LMs, impacting training stability and model performance.

The fourth stage focuses on privacy reduction, acknowledging concerns related to privacy breaches when using web-based data. Privacy reduction measures are implemented to remove personally identifiable information (PII) from the pre-training corpus. The final step, tokenization, segments raw text into individual tokens, which are then input into LLMs.

Following pre-training, the LLM undergoes fine-tuning, involving training on a specific task or domain. Labeled examples guide the model to generate more accurate and contextually appropriate responses for the target task. Fine-tuning enables LLMs to specialize in applications like language translation, question answering, or text generation.

LLMs excel due to their ability to capture statistical patterns and linguistic nuances in the training data. By processing vast amounts of text, they develop a comprehensive understanding of language, producing coherent and contextually relevant responses.

During the inference stage, when interacting with an LLM, a user inputs a prompt or query, and the model generates a response based on learned knowledge and context. This response is generated using probabilistic methods that consider the likelihood of various words or phrases given the input context, showcasing the data preprocessing pipelines for pre-training LLMs.

Tokenization

In the training of LLMs to predict text, a fundamental preprocessing step is tokenization, a common practice in natural language processing systems. Tokenization aims to break down the text into nondecomposing units known as tokens, with these tokens being characters, subwords, symbols, or words depending on the model's size and type.

- **WordPiece**: Initially introduced as a novel text segmentation technique for Japanese and Korean languages to enhance language models for voice search systems, WordPiece selects tokens that increase the likelihood of an n-gram-based language model trained on a vocabulary composed of tokens.

- **BPE**: Originating from compression algorithms, **byte pair encoding (BPE)** is an iterative process that generates tokens by replacing pairs of adjacent symbols with a new symbol. This approach involves merging occurrences of the most frequently appearing symbols in the input text.

- **UnigramLM**: In this tokenization method, a basic unigram language model (LM) is trained using an initial vocabulary of subword units. The vocabulary undergoes iterative pruning by removing the least probable items from the list, identified as the underperforming elements in the unigram LM.

Attention

The concept of attention, particularly selective attention, has been extensively examined within the realms of perception, psychophysics, and psychology. Selective attention can be understood as *"the programming by the O of which stimuli will be processed or encoded and in what order this will occur."*

CHAPTER 2 WHAT ARE LARGE LANGUAGE MODELS?

While this definition originates from visual perception, it bears striking resemblances to the recently formulated attention mechanisms[7] (determining which stimuli will be processed) and positional encoding (deciding the order of processing) in LLMs.

Attention Mechanisms in LLMs

The attention mechanism plays a crucial role in computing a representation of input sequences by establishing relationships between different positions (tokens) within these sequences. Various approaches exist for calculating and implementing attention, and some well-known types are outlined here:

- **Self-attention:** Also referred to as intra-attention, self-attention involves all queries, keys, and values originating from the same block (either encoder or decoder). This layer establishes connections between all sequence positions with $O(1)$ space complexity, making it highly effective for learning long-range dependencies in the input.

- **Cross-Attention:** In encoder-decoder architectures, the outputs of encoder blocks serve as queries for the intermediate representation of the decoder. This setup provides keys and values to compute a representation of the decoder conditioned on the encoder, and it is termed cross-attention.

- **Full Attention:** The straightforward implementation of self-attention is known as full attention.

- **Sparse Attention:** Self-attention has a time complexity of $O(n^2)$, which becomes impractical when scaling LLMs to handle large context windows. An approximation to self-attention was proposed, significantly improving the ability of GPT series LLMs to process a greater number of input tokens within a reasonable timeframe.

---

[7] S. Biderman, H. Schoelkopf, Q. Anthony, H. Bradley, K. O'Brien, E. Hallahan, M. A. Khan, S. Purohit, U. S. Prashanth, E. Raff, et al., "Pythia: A suite for analyzing large language models across training and scaling," arXiv preprint arXiv:2304.01373, 2023

- **Flash Attention:** The bottleneck in calculating attention using GPUs lies in memory access rather than computational speed. Flash Attention utilizes the classical input tiling approach to process input blocks in GPU on-chip SRAM instead of performing I/O for each token from High Bandwidth Memory (HBM). Extending this approach to sparse attention replicates the speed gains seen in full attention implementation. This innovation allows LLMs to handle even larger context-length windows compared to those using sparse attention.

Positional Encoding

The attention modules, as designed, do not inherently account for the order of processing. To address this, transformers introduced "positional encodings" to incorporate information about the positions of tokens in input sequences. Various positional encoding variants have been proposed. Intriguingly, a recent study[8] suggests that incorporating this information may not significantly impact state-of-the-art decoder-only transformers.

- **Absolute:** The most direct method involves adding sequence order information by assigning a unique identifier to each position in the sequence before presenting it to the attention module.

- **Relative:** To convey information about the relative dependencies among tokens appearing at different locations in the sequence, a relative positional encoding is computed through some form of learning.

Two notable types of relative encodings are as follows:

- **Alibi:** This approach subtracts a scalar bias from the attention score calculated using two tokens, and this bias increases with the distance between the positions of the tokens. This learned method effectively favors recent tokens for attention.

---

[8] M. Irfan, A. I. Sanka, Z. Ullah, R. C. Cheung, "Reconfigurable content-addressable memory (CAM) on FPGAs: A tutorial and survey," Future Generation Computer Systems, vol. 128, pp. 451–465, 2022

- **RoPE:** In LLMs, keys, queries, and values are all vectors. RoPE involves rotating the query and key representations at an angle proportional to their absolute positions in the input sequence. This rotation results in a relative positional encoding scheme that diminishes with the distance between the tokens.

Activation Functions

Activation functions play a crucial role in enhancing the curve-fitting capabilities of neural networks. The contemporary activation functions employed in LLMs differ from earlier squashing functions but are integral to the success of LLMs.

ReLU

The Rectified Linear Unit (ReLU) is defined as `ReLU(x) = max(0, x) (1)`.

GELU

The Gaussian Error Linear Unit (GELU) combines ReLU, dropout, and zoneout. It stands out as the most widely utilized activation function in current LLM literature.

GLU Variants

The Gated Linear Unit is a neural network layer involving an element-wise product (⊗) of a linear transformation and a sigmoid-transformed (σ) linear projection of the input, given by the equation:

`GLU(x,W,V,b,c)=(xW+b)⊗σ(xV+c) (2)`

Here, x represents the input of the layer, and W, b, V, and c are learned parameters.

GLU was modified to assess the impact of various variations in the training and testing of transformers, leading to improved empirical results. The following are different GLU variations introduced[9] and utilized in LLMs:

---

[9] C. Raffel, N. Shazeer, A. Roberts, K. Lee, S. Narang, M. Matena, Y. Zhou, W. Li, and P. J. Liu, "Exploring the limits of transfer learning with a unified text-to-text transformer," The Journal of Machine Learning Research, vol. 21, no. 1, pp. 5485–5551, 2020

- ReGLU(x,W,V,b,c)=max(0,xW+b)⊗
- GEGLU(x,W,V,b,c)=GELU(xW+b)⊗(xV+c)
- SwiGLU(x,W,V,b,c,β)=Swishβ(xW+b)⊗(xV+c)

Layer Normalization

Layer normalization facilitates quicker convergence and is a prevalent component in transformers. In this section, we explore various normalization techniques extensively employed in LLM literature.

LayerNorm

Layer normalization calculates statistics across all the hidden units within a layer (l) using the following formula:

$$u^l = \frac{1}{n}\sum_{i}^{n} a_i^l \quad \sigma^l = \sqrt{\frac{1}{n}\sum_{i}^{n}\left(a_i^l - u^l\right)^2} n$$

In this equation, where n represents the number of neurons in layer l and a_i^l is the aggregated input of the *i-th* neuron in layer l, LayerNorm offers invariance to both weight rescaling and distribution re-centering.

RMSNorm

RMSNorm challenges the presumed invariance properties of LayerNorm. It suggests that comparable performance benefits to those of LayerNorm can be achieved through a computationally efficient normalization technique that compromises re-centering invariance for speed. The normalized summed input to layer \(l\) in LayerNorm is expressed as follows:

$$\overline{a_i^l} = \frac{a_i^l - u^l}{\sigma} g_i^l$$

Pre-norm and Post-norm

In the realm of large language models (LLMs), the transformer architecture serves as the foundation with certain variations. The original implementation incorporated layer normalization after the residual connection, commonly known as post-LN. This sequence follows the order of Multi-head attention–Residual–LN. An alternative sequence, referred to as pre-LN, involves placing the normalization step before the self-attention layer, following the order of LN–Multi-head attention–Residual. Pre-LN is recognized for providing increased stability in training.

DeepNorm

Despite the advantages of pre-LN over post-LN training, pre-LN training can inadvertently impact gradients,[10] leading to larger gradients in the earlier layers compared to those at the bottom. DeepNorm is introduced as a solution to mitigate these adverse effects on gradients. Its formulation is expressed as

$$x^{l_f} = LN(\alpha x^{l_p} + G^{l_p}(x^{l_p}, \theta^{l_p}))$$

In this expression, α denotes a constant, and θ_lp represents the parameters of layer lp. These parameters undergo scaling by another constant β. Both of these constants are architecture dependent.

Distributed Training of Large Language Models (LLMs)

Distributed training of large language models (LLMs) involves leveraging multiple computing resources simultaneously to efficiently process and learn from vast datasets. It accelerates training times and enables the handling of models with billions of parameters, ensuring scalability and enhanced performance. By distributing the workload, it becomes feasible to train complex models that would otherwise be computationally prohibitive.

---

[10] J. Hoffmann, S. Borgeaud, A. Mensch, E. Buchatskaya, T. Cai, E. Rutherford, D. d. L. Casas, L. A. Hendricks, J. Welbl, A. Clark, et al., "Training compute-optimal large language models," arXiv preprint arXiv:2203.15556, 2022

There are various distributed training approaches for LLMs:

- **Data Parallelism:** This approach involves replicating the model across multiple devices, with batch data divided among these devices. At the conclusion of each training iteration, weights are synchronized across all devices.

- **Tensor Parallelism:** Tensor parallelism distributes tensor computations across devices, also known as horizontal parallelism or intra-layer model parallelism.

- **Pipeline Parallelism:** In pipeline parallelism, model layers are distributed across different devices, also termed vertical parallelism.

- **Model Parallelism:** Model parallelism is a combination of tensor and pipeline parallelism.

- **3D Parallelism:** This approach integrates data, tensor, and model parallelism.

- **Optimizer Parallelism:** Also known as zero redundancy optimizer, optimizer parallelism implements optimizer state partitioning, gradient partitioning, and parameter partitioning across devices. This reduces memory consumption while minimizing communication costs.

Data Preprocessing

Data preprocessing is a crucial step in machine and deep learning that involves cleaning, transforming, and organizing raw data into a usable format. This process enhances data quality, ensures consistency, and prepares the dataset for effective analysis and model training, ultimately leading to more accurate and reliable results.

The data preprocessing techniques utilized in the training of large language models (LLMs) include the following:

- **Quality Filtering:** Ensuring the quality of training data is crucial for optimal results. Two approaches to filter data are employed: (1) classifier based and (2) heuristics based. Classifier-based methods train a classifier on high-quality data to predict the text's quality for filtering. On the other hand, heuristics-based approaches utilize predefined rules, including language, metrics, statistics, and keywords, for data filtering.

- **Data Deduplication:** Duplicated data can negatively impact model performance and lead to data memorization. Therefore, data deduplication is a crucial preprocessing step in training LLMs. This process can be applied at various levels, such as sentences, documents, and entire datasets.

- **Privacy Reduction:** The majority of training data for LLMs is sourced from the web, containing private information. To address privacy concerns, many LLMs employ heuristics-based methods to filter out sensitive information like names, addresses, and phone numbers, preventing the model from learning personal details.

Architectures

In this section, different variants of transformer architectures are explored at a higher level, which stem from variations in attention application and the interconnection of transformer blocks. Figure 2-2 illustrates the attention patterns of these architectures.

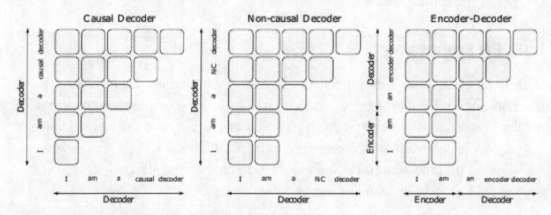

Figure 2-2. Causal decoder vs. non-causal decoder vs. encoder-decoder architectures (Source: https://arxiv.org/)

Encoder-Decoder

Originally designed for sequence transduction models, transformers adopted the encoder-decoder architecture, particularly for tasks like machine translation. This architectural design involves an encoder that encodes input sequences into variable-length context vectors. These vectors are then passed to the decoder, aiming to minimize the discrepancy between predicted token labels and the actual target token labels.

Causal Decoder

The primary goal of an LLM is to predict the next token based on the input sequence. While the encoder provides additional context, it has been observed that LLMs can perform well even without an encoder, relying solely on the decoder. Similar to the original encoder-decoder architecture's decoder block, this decoder limits information flow backward. In other words, the predicted token tk depends only on the tokens preceding it, up to tk−1. This variant is widely utilized in state-of-the-art LLMs.

Prefix Decoder

Causal masked attention is justifiable in encoder-decoder architectures, where the encoder can attend to all tokens in the sentence from any position using self-attention. However, when the encoder is removed, and only the decoder is retained, this flexibility in attention is lost. A variation in decoder-only architectures involves changing the mask from strictly causal to fully visible on a portion of the input sequence. The Prefix decoder is also referred to as the non-causal decoder architecture.

Pre-training Objectives

The pre-training objectives of large language models (LLMs) include the following:

- **Full Language Modeling:** This entails an autoregressive language modeling objective where the model is tasked with predicting future tokens based on preceding tokens.
- **Prefix Language Modeling:** Introducing a non-causal training objective, this approach involves randomly selecting a prefix, and the loss is calculated based on the remaining target tokens.

- **Masked Language Modeling:** In this training objective, tokens or spans (a sequence of tokens) are randomly masked, and the model is required to predict the masked tokens considering the past and future context.

- **Unified Language Modeling:** Unified language modeling combines causal, non-causal, and masked language training objectives. In masked language modeling within this framework, attention is unidirectional, flowing either from left-to-right or right-to-left context.

Model Adaptation

This section provides an overview of the fundamental stages of large language models (LLMs) adaptation, spanning from pre-training to fine-tuning for downstream tasks and practical application. The term "alignment-tuning" is used to denote aligning with human preferences, while the literature may occasionally employ the term "alignment" for different purposes.

Pre-training

In the initial stage, the model undergoes self-supervised training on a large corpus, predicting the next tokens based on the input. The design choices for LLMs encompass a range of architectures, including encoder-decoder and decoder-only, employing diverse building blocks and loss functions.

Fine-Tuning

Fine-tuning LLMs can be achieved through various approaches:

- **Transfer Learning:** Pre-trained LLMs demonstrate robust performance across various tasks. However, to enhance performance for a specific downstream task, pre-trained models undergo fine-tuning with task-specific data, a process known as transfer learning.

- **Instruction-Tuning:** To enable effective responses to user queries, pre-trained models are fine-tuned on instruction-formatted data, consisting of instructions and corresponding input-output pairs.

Instructions typically encompass multitask data in plain natural language, guiding the model to respond appropriately to the prompt and input. This fine-tuning strategy improves zero-shot generalization and downstream task performance.

- **Alignment-Tuning:** LLMs may generate false, biased, or harmful text. To rectify this, models are aligned using human feedback to ensure helpful, honest, and harmless outputs. Alignment-tuning involves prompting LLMs to generate unexpected responses, and their parameters are then updated to avoid such responses.[11]

Alignment Verification and Utilization

Ensuring that large language models (LLMs) align with human intentions and values is crucial. A model is considered "aligned" if it satisfies the three criteria of being helpful, honest, and harmless (HHH).

Researchers employ Reinforcement Learning with Human Feedback (RLHF) for model alignment. In RLHF, a fine-tuned model on demonstrations undergoes further training with reward modeling (RM) and reinforcement learning (RL). The following provides a brief discussion of the RM and RL pipelines in RLHF.

- **Reward Modeling:** This process trains a model to rank generated responses based on human preferences using a classification objective. Humans annotate responses generated by LLMs, providing feedback based on the HHH criteria.

- **Reinforcement Learning:** Combined with the reward model, reinforcement learning is used for alignment in the subsequent stage. The previously trained reward model ranks LLM-generated responses as preferred or dispreferred, aiding in aligning the model with proximal policy optimization (PPO). This iterative process continues until convergence.

---

[11] "An era of ChatGPT as a significant futuristic support tool: A study on features, abilities, and challenges," BenchCouncil Transactions on Benchmarks, Standards and Evaluations, vol. 2, no. 4, 100089, 2022

Prompting/Utilization

Prompting is a method for querying trained LLMs to generate responses. LLMs can be prompted in various setups, adapting to instructions either without fine-tuning or with fine-tuning on data containing different prompt styles:

- **Zero-Shot Prompting:**[12] LLMs exhibit zero-shot learning capabilities, answering queries without having seen any examples in the prompt.

- **In-Context Learning:**[13] Also known as few-shot learning, this approach involves presenting multiple input-output demonstration pairs to the model to generate the desired response.

- **Reasoning in LLMs:**[14] LLMs can act as zero-shot reasoners, providing answers to logical problems, task planning, critical thinking, etc., with reasoning. Different prompting styles are used to provoke reasoning, and methods train LLMs on reasoning datasets to enhance their performance. Various prompting techniques for reasoning include the following:

  - **Chain-of-Thought (CoT):**[15] A special prompting case where demonstrations include reasoning information aggregated with inputs and outputs, guiding the model to generate outcomes with step-by-step reasoning.

    - **Self-consistency:** Improves CoT performance by generating multiple responses and selecting the most frequent answer

    - **Tree of Thought (ToT):** Explores multiple reasoning paths with possibilities to look ahead and backtrack for problem-solving

---

[12] Adrian Tam, What Are Zero-Shot Prompting and Few-Shot Prompting, https://machinelearningmastery.com/what-are-zero-shot-prompting-and-few-shot-prompting/

[13] Qingxiu Dong, Lei Li, Damai Dai, Ce Zheng, Jingyuan Ma, Rui Li, Heming Xia, Jingjing Xu, Zhiyong Wu, Baobao Chang, Xu Sun, Lei Li, Zhifang Sui, A Survey on In-context Learning, https://arxiv.org/abs/2301.00234

[14] Jie Huang and Kevin Chen-Chuan Chang, Towards Reasoning in Large Language Models: A Survey, https://arxiv.org/abs/2212.10403

[15] Jason Wei, Xuezhi Wang, Dale Schuurmans, Maarten Bosma, Brian Ichter, Fei Xia, Ed Chi, Quoc Le, Denny Zhou, Chain-of-Thought Prompting Elicits Reasoning in Large Language Models, https://arxiv.org/abs/2201.11903

- **Single-Turn Instructions:**[16] In this setup, LLMs are queried only once, with all relevant information in the prompt. LLMs generate responses by understanding the context, either in a zero-shot or few-shot setting.

- **Multi-turn Instructions:**[17] Solving complex tasks requires multiple interactions with LLMs, where feedback and responses from other tools are given as input to the LLM for subsequent rounds. This style of using LLMs in the loop is common in autonomous agents.

Training of LLMs

Developing large language models encompasses several essential steps crucial for their successful training. The process typically initiates with the gathering and preprocessing of an extensive volume of text data from diverse sources, including books, articles, websites, and various textual corpora. The meticulously curated dataset forms the cornerstone for the training of large language models (LLMs). The training procedure itself revolves around unsupervised learning, wherein the model acquires the ability to predict the subsequent word in a sequence based on the preceding context. This particular task is commonly known as language modeling.

LLMs employ advanced neural network architectures, such as transformers, enabling them to capture intricate patterns and dependencies within language. The primary training objective is to optimize the model's parameters, aiming to maximize the likelihood of generating the correct next word within a given context. This optimization is typically achieved through an algorithm known as stochastic gradient descent (SGD) or its variants, coupled with backpropagation, which iteratively computes gradients to update the model's parameters.

---

[16] Humza Naveeda, Asad Ullah Khana, Shi Qiub, Muhammad Saqibc, Saeed Anwar, Muhammad Usman, Naveed Akhtarg, Nick Barnesh, Ajmal Miani, A Comprehensive Overview of Large Language Models

[17] Yuchong Sun, Che Liu, Kun Zhou, Jinwen Huang, Ruihua Song, Wayne Xin Zhao, Fuzheng Zhang, Di Zhang, Kun Gai, Parrot: Enhancing Multi-Turn Instruction Following for Large Language Models

Benefits and Challenges of LLMs in Various Domains

LLMs have a broad spectrum of applications, and this discussion focuses on their impactful roles in medicine, education, finance, and engineering. These fields were selected due to their significant influence and the transformative potential of LLMs within these domains. The versatility and capability of LLMs to tackle complex challenges, like the ones in Figure 2-3, and aid human efforts are well demonstrated through these applications.

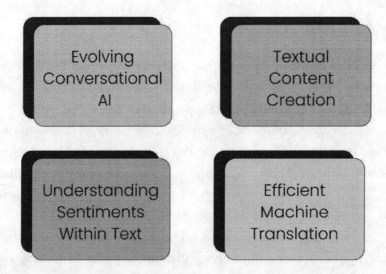

Figure 2-3. Top four benefits of large language models

General Purpose

LLMs stand out as versatile tools capable of handling a diverse array of tasks, even those beyond their specific training. Their proficiency lies in their capacity to comprehend, produce, and modify text in ways that are contextually appropriate and human-like. This versatility enables them to undertake a range of functions, from straightforward language translation and answering questions to more intricate activities like text summarization, creative writing, and programming assistance. LLMs' adaptability extends to mimicking the style and tone of the text they are working with, resulting in outputs that are both user-centric and context-sensitive.

In everyday scenarios, LLMs find utility as virtual personal assistants, aiding in tasks such as composing emails or organizing meetings. They are also increasingly employed in customer service, handling routine inquiries and thereby allowing human staff to focus on more complex matters. Additionally, LLMs are being used in content creation for digital platforms, generating human-like text based on specified prompts. Another significant role of LLMs is in data analysis, where they can sift through and summarize large volumes of text data, identifying patterns and key points more rapidly than human analysts.

Despite their broad utility, it's crucial to acknowledge that LLMs, like any AI technology, are limited by the quality of their training data. As such, they need to be utilized cautiously, given that they may inadvertently reflect biases present in their training data, leading to potentially skewed or inaccurate outcomes.

Medical Applications

In healthcare, LLMs such as ChatGPT have shown remarkable promise across various medical applications. They have been effectively used in medical education, radiology decision-making, clinical genetics, and patient care, as evidenced by numerous studies. In medical education, for instance, ChatGPT has become a valuable interactive learning and problem-solving tool. Its performance on the United States Medical Licensing Exam (USMLE) has been notably proficient, achieving scores that meet or surpass the passing criteria, showcasing its deep understanding of medical knowledge without needing specific training.

Studies point toward the future development of AI-driven clinical decision-making tools. ChatGPT, for instance, has shown potential in radiological decision-making, enhancing clinical workflows and the responsible use of radiology services. Research by Kung and others has indicated that LLMs like ChatGPT could significantly improve personalized, compassionate, and scalable healthcare delivery, aiding both in education and clinical decision-making.

A study on clinical genetics found ChatGPT's performance comparable to human responses in answering genetics-related questions, excelling in memory-based queries but less so in critical thinking tasks. This research also noted ChatGPT's ability to provide varied, plausible explanations for both correct and incorrect responses. Additionally, research evaluating ChatGPT's accuracy in life support and resuscitation questions found it to be largely accurate, especially in questions from the American Heart Association's Basic Life Support and Advanced Cardiovascular Life Support exams.

In neurosurgical research and patient care, the potential roles of ChatGPT have been explored in patient data collection, survey administration, and providing care and treatment information. However, the deployment of such technologies requires careful consideration to ensure their effectiveness and safety. A comprehensive study encompassing a range of AI applications in life sciences, including decision support, NLP, data mining, and machine learning, emphasizes the importance of reproducibility in AI model development and addresses ongoing research challenges.

AI-driven chatbots like ChatGPT are poised to improve patient outcomes by facilitating better communication between patients and healthcare providers. Using NLP, these chatbots can offer patients easily understandable information about their care and treatment. Additionally, AI is being utilized in developing databases for COVID-19 drug repurposing, and existing tools like Ada Health, Babylon Health, and Buoy Health already facilitate patient interaction. The rise of LLMs could further boost patient confidence in these chatbots and enhance the services they offer. For example, tools like XrayGPT are being developed for automated analysis of X-ray images, allowing patients to engage in interactive dialogues about their condition.

In the realm of artificial intelligence, large language models (LLMs) like GPT-4 have revolutionized how machines understand and generate human language. Their capabilities extend beyond mere text generation, impacting various sectors including technology, education, business, healthcare, and creative arts.

Healthcare Communication and Management

LLMs are revolutionizing healthcare communication. They assist in patient interaction, providing clear explanations of medical conditions and treatments. LLMs can also help in drafting patient information leaflets, ensuring they are understandable and accessible. In hospital management, they assist in organizing patient records, scheduling, and handling administrative tasks, thus improving overall efficiency.

Enhanced Natural Language Processing

LLMs are the backbone of advanced natural language processing (NLP). They excel in understanding context, generating coherent responses, and improving communication interfaces. This is evident in chatbots, virtual assistants, and customer service applications where LLMs provide more natural, human-like interactions. For instance, they can handle complex customer queries, offer personalized recommendations, and automate routine tasks, significantly improving efficiency and user experience.

Education

Recent discussions have highlighted the transformative role of artificial intelligence (AI) in education, particularly concerning student assignments and examinations. Since OpenAI introduced ChatGPT, there has been a noticeable shift in how students engage with educational content, assignments, and coursework.

One significant benefit of incorporating ChatGPT and AI bots in educational settings is the enhancement of assignment completion efficiency. ChatGPT, for instance, is adept at producing high-quality answers across diverse prompts, thereby saving students considerable time and effort in their academic work. Furthermore, AI bots have the potential to streamline the grading process, lightening the load on educators while affording them the opportunity to offer more comprehensive feedback to students.

Another key advantage of these AI tools is their ability to deliver tailored learning experiences. AI bots can assess a student's performance on past assignments and tests, using this data to tailor future learning recommendations. This approach helps students identify their academic strengths and weaknesses, allowing them to concentrate on areas needing improvement.

Khan Academy, a renowned educational non-profit, has expressed interest in leveraging ChatGPT through its AI chatbot, Khanmigo. This virtual tutor and classroom assistant aims to enrich tutoring and coaching by facilitating direct interactions with students. Such initiatives reflect the optimism surrounding AI's potential in education, challenging the misconception that its primary use is for cheating. AI technology, though still evolving, is seen as promising for catering to diverse student needs.

Nevertheless, the use of ChatGPT and AI bots in education is not without potential downsides. A primary concern is the risk of diminishing creativity and critical thinking skills among students. Overreliance on AI for assignments and exams might hinder the development of essential skills needed for independent problem-solving and critical analysis.

Content Creation and Augmentation

One of the most prominent uses of LLMs is in content creation. They assist in generating articles, reports, essays, and creative writing, reducing the time and effort required for these tasks. LLMs like GPT-4 can produce content on a wide range of topics, adhering to specified styles and guidelines. Moreover, they assist in editing and refining text, enhancing quality and coherence.

Language Translation and Localization

LLMs have made significant strides in breaking language barriers. They offer real-time, accurate translation services, enabling seamless communication across different languages. This capability is crucial for global businesses, travel, and cross-cultural interactions. Additionally, LLMs assist in localization, ensuring that content is culturally appropriate and resonates with local audiences.

Research and Data Analysis

Researchers utilize LLMs for literature reviews, hypothesis generation, and data interpretation. These models can process vast amounts of information rapidly, identifying patterns and insights that might be missed by human researchers. This capability is invaluable in fields like medicine, where LLMs can assist in diagnosing diseases or suggesting treatments based on medical literature.

Finance

Large language models (LLMs) are increasingly influential in the finance sector, offering a range of applications from financial natural language processing (NLP) tasks to risk analysis, algorithmic trading, market forecasting, and financial reporting. Models like BloombergGPT, a 50-billion-parameter LLM trained on a vast and diverse financial dataset, have significantly improved tasks like news categorization, entity identification, and query answering in financial NLP. Leveraging the extensive financial data at its disposal, this model significantly enhances customer service by effectively addressing customer inquiries and providing top-tier financial advice.

LLMs are also instrumental in risk management and assessment. By examining historical market trends and data, these models can pinpoint potential risks and suggest mitigation strategies through various financial algorithms. Financial entities are employing these models for more informed decision-making in areas such as credit risk evaluation, loan sanctioning, and investment planning. Moreover, LLMs are finding use in algorithmic trading, where they apply their predictive analytics to spot trading opportunities in the market.

Nevertheless, the sensitive nature of financial data and related privacy concerns necessitate measures like data encryption, redaction, and adherence to data protection policies to ensure LLMs operate in line with data privacy standards. A recent initiative in this context is FinGPT, an open source LLM designed specifically for finance, indicating a growing interest and ongoing developments in this area.

Creative Arts

In the realm of creative arts, LLMs are tools for inspiration and creativity. They assist artists, writers, and musicians in generating ideas, lyrics, scripts, and even entire compositions. This collaborative process between AI and human creativity is spawning new forms of art and entertainment.

Ethical and Responsible Use

Despite these advantages, the use of LLMs raises concerns regarding ethics, bias, and misinformation. Ensuring responsible and ethical use is paramount. This involves continuous monitoring, updating models to reduce biases, and implementing guidelines to prevent misuse.

Legal and Compliance Assistance

In the legal sector, LLMs are increasingly being utilized for the thematic examination of legal documents. They play a pivotal role in the initial coding of datasets, identifying key themes, and categorizing data based on these themes. The synergy between legal professionals and LLMs has been particularly beneficial in the analysis of legal texts, such as judicial opinions on theft cases, enhancing both the efficiency and the quality of legal research. Furthermore, LLMs have shown proficiency in providing explanations for legal terminology, with a focus on enhancing factual accuracy and relevance. This is achieved by integrating sentences from relevant case law into the LLM, enabling it to produce more accurate and high-quality explanations with fewer factual errors. Additionally, LLMs have been developed and trained with specific legal knowledge, enabling them to engage in legal reasoning tasks and respond to legal queries effectively.

Financial Analysis and Forecasting

In finance, LLMs are used for market analysis, forecasting, and risk assessment. They can analyze financial reports, news, and market data to provide insights into investment opportunities or economic trends. This helps investors and companies make informed decisions. Also, they assist in automating routine financial tasks like report generation and data analysis.

Disaster Response and Management

In disaster response, LLMs can play a crucial role in analyzing data from various sources to provide real-time information about affected areas. They can assist in drafting emergency communication, coordinating response efforts, and managing logistics. This can be critical in saving lives and mitigating disaster impacts.

Personalized Marketing and Customer Insights

LLMs offer personalized marketing solutions by analyzing customer data and generating targeted content. They help in creating personalized email campaigns, social media content, and advertising copy that resonates with specific audience segments. Additionally, they can analyze customer feedback and social media discussions to provide insights into consumer behavior and preferences.

Gaming and Interactive Entertainment

In the gaming industry, LLMs are used to create dynamic, responsive narratives and dialogues. They can generate storylines, characters, and dialogues that adapt to player choices, creating a more immersive gaming experience. They also assist in game design by generating ideas and content for game worlds and quests.

Accessibility Enhancements

LLMs significantly contribute to making technology more accessible. They can generate real-time captions and descriptions for audio and video content, aiding those with hearing or visual impairments. They also improve voice recognition software, making technology more accessible for individuals with different speech patterns or accents.

Environmental Monitoring and Sustainability

In environmental science, LLMs assist in analyzing data from various sources to monitor climate change and environmental degradation. They can help in drafting reports on environmental impact and suggest sustainable practices. This is vital for research and policy-making in environmental conservation.

LLMs and Engineering Applications

LLMs are increasingly recognized for their wide-ranging impact in various engineering disciplines. In software engineering, for instance, ChatGPT has been applied to tasks like code generation, debugging, software testing, NLP, documentation creation, and team collaboration. It aids developers in producing code snippets, identifying and rectifying errors, formulating test scenarios, and streamlining software documentation and teamwork processes. The language comprehension and generation skills of ChatGPT improve efficiency and communication in software engineering.

Specifically, in software engineering, ChatGPT enables the generation of code from natural language descriptions, saving developers time and enhancing productivity. It also aids in debugging by identifying errors and proposing solutions, thereby expediting the debugging process. For software testing, ChatGPT can create test cases from natural language inputs, boosting the effectiveness of testing processes.

In mechanical engineering, a study by Tiro explored the use of ChatGPT for mechanical calculations but found limitations in accuracy, leading to the discontinuation of this research avenue. This suggests that, in its current state, ChatGPT may not be reliable for practical engineering problem-solving and should be used cautiously.

In mathematics education, studies like those by Wardat et al. indicate that ChatGPT can assist in teaching mathematics, providing interactive and tailored learning experiences. As a virtual tutor, it offers real-time feedback and custom problem-solving strategies. However, the model's effectiveness in solving mathematical problems can vary based on the complexity and the accuracy of the inputs provided. Frieder et al.'s research further investigates ChatGPT's mathematical capabilities, comparing it with other models and assessing its usefulness to professional mathematicians.

In manufacturing, Wang et al.'s study evaluates ChatGPT's utility in supporting design and manufacturing education. While the model shows promise in generating coherent content and initial solutions, it sometimes struggles with understanding queries and applying knowledge accurately. Conversely, Badini et al.'s study in additive manufacturing troubleshooting found ChatGPT to be remarkably accurate and organized in its approach, particularly in addressing specific technical issues in Fused Filament Fabrication. They suggest integrating ChatGPT into Additive Manufacturing software for real-time optimization, potentially enhancing process efficiency and quality.

Overall, while LLMs like ChatGPT show significant potential in various engineering domains, their current limitations underline the importance of using them as supplementary tools alongside traditional methods and human oversight.

Chatbots

Chatbots are increasingly common in customer service roles, adept at handling inquiries, providing support, and resolving issues. Their applications extend to entertainment, healthcare, and education sectors. These chatbots are often integrated with LLMs to craft more advanced and interactive conversational experiences. For instance, a chatbot might employ an LLM like ChatGPT to enhance the quality of its textual responses. Well-known examples of such chatbots include ChatGPT, Google Bard, and Microsoft Bing. An illustration of an educational interaction with ChatGPT is provided as an example.

In terms of comparison, ChatGPT and Google Bard represent two of the leading LLMs currently in use. Both are skilled in generating text, translating languages, creating various forms of creative content, and providing informative responses to queries. Despite their similarities, notable differences exist between these models. For instance, ChatGPT is recognized for its creative capabilities, whereas Google Bard is known for its authenticity. A detailed comparison of ChatGPT, Google Bard, and Microsoft Bing Chatbots is presented, highlighting the unique features and capabilities of each.

LLM Agents

LLM agents are advanced AI systems built using large-scale language models like OpenAI's GPT-4. Think of them as highly intelligent virtual assistants that can understand and generate human language, making them incredibly versatile for various tasks.

LLM agents can grasp the meaning behind complex questions and instructions, making them great at comprehending human language. They can write like humans, drafting emails, creating reports, and even writing stories or articles.

These agents can chat with you, providing relevant responses and engaging in meaningful dialogues. Perfect for customer service or virtual assistants. They can pull information from vast databases and even use real-time data to give you up-to-date answers.

LLM agents are also scalable. They can handle a ton of interactions at the same time, making them ideal for businesses with high customer engagement.

They are able to take care of repetitive tasks, freeing up human workers for more complex duties. By automating routine tasks, they can help save on operational costs, which makes them cost-effective.

LLM Limitations

While large language models (LLMs) have contributed significantly to natural language processing, they are accompanied by a range of shortcomings. This section outlines various such limitations. These include biases in the data used for training, an overdependence on superficial patterns, a lack of robust common sense, challenges in reasoning and processing feedback, a need for extensive data and computing power, and issues with generalizing their knowledge.

Further, LLMs struggle with interpretability, handling rare or unknown words, grasping syntax and grammar intricacies, possessing domain-specific expertise, and vulnerability to targeted misinformation attempts. Ethical concerns are also prominent, along with difficulties in contextual language processing, emotion and sentiment analysis, multilingual support, and memory constraints. Their creativity is limited, as are their real-time processing, training, and maintenance affordability. Scalability, causal understanding, multimodal input processing, attention span, transfer learning, world knowledge beyond text, and human behavior and psychology comprehension are additional limitations.

CHAPTER 2   WHAT ARE LARGE LANGUAGE MODELS?

LLMs also face challenges in producing extended text, collaborating effectively, managing ambiguous information, appreciating cultural nuances, learning incrementally, processing structured data, and dealing with noisy or erroneous input. Given these numerous limitations, it is crucial for researchers and practitioners to recognize and address these challenges to ensure responsible and effective LLM usage, and to strive for the development of new models that overcome these constraints.

Bias

Bias in language models arises when their training data reflects existing prejudices. As highlighted by Schramowski[18] and colleagues, these models, although aimed at emulating natural language, can inadvertently propagate biases, leading to unfair or skewed outputs. This can spark critique in various sectors such as politics, law, and society. The forms of bias include the following:

- **Training Data Bias:** These models are trained on vast human language datasets. If these datasets carry biases regarding race, gender, or socioeconomic status, the model might echo these biases. For instance, gender-biased data could result in the model favoring one gender over another in its outputs.

- **User Interaction Bias:** The input from users shapes the responses of chatbots. If the input is consistently biased or prejudiced, the model may learn and repeat these biases. For example, a model exposed to frequent discriminatory queries against a certain group might start to reflect these biases in its responses.

- **Algorithmic Bias:** The algorithms used for training and operating these models can introduce bias. If a model is trained to optimize specific metrics like accuracy or user engagement, it might produce biased responses that align with these metrics.

---

[18] P. Schramowski, C. Turan, N. Andersen, C. A. Rothkopf, K. Kersting, "Large pre-trained language models contain human-like biases of what is right and wrong to do," Nature Machine Intelligence, vol. 4, no. 3, pp. 258–268, 2022

- **Contextual Bias:** Chatbots respond based on the context they receive. If this context includes biases related to the user's location, language, or other factors, the responses may be biased. An example is a model generating biased responses about a culture or religion if it lacks adequate information in that specific context.

Understanding and addressing these biases is crucial for ensuring fairness and accuracy in language model outputs.

Hallucinations

LLMs occasionally produce content that deviates from factual accuracy, a phenomenon known as "information hallucination."[19] This typically occurs when the model attempts to bridge gaps in its knowledge or context, drawing upon learned patterns rather than actual data. Such instances can result in false or misleading information, which is especially concerning in critical applications.

The underlying reasons for these hallucinations are the subject of ongoing investigation. Current research indicates that the issue may stem from various aspects, including the training process, the dataset used, and the model's structure. LLMs may have a tendency to generate more engaging or coherent content, inadvertently increasing the likelihood of hallucinations.

Efforts to curb hallucinations have led to several strategies. One such method involves adjusting the training regimen to discourage hallucinations, as seen in techniques like "reality grounding." Expanding the training dataset to be more varied and extensive could also diminish the model's tendency to make unfounded assumptions.

Another avenue being explored is training models with data that is verifiable or capable of being fact-checked. This approach aims to make the model more reliant on factual information over assumptions. However, implementing this strategy requires a meticulous selection of data and metrics.

---

[19] T. McCoy, E. Pavlick, and T. Linzen, "Right for the wrong reasons: Diagnosing syntactic heuristics in natural language inference," in Proceedings of the 57th Annual Meeting of the Association for Computational Linguistics (Florence, Italy), pp. 3428–3448, Association for Computational Linguistics, July 2019

Future research is crucial for a deeper understanding and better management of hallucinations in LLMs. This may include developing more advanced models capable of distinguishing between fact and assumption, as well as innovating training methodologies and datasets.

Vulnerability to Various Types of Cyber Attacks

The model is prone to different forms of adversarial attacks. These include "prompt injection," where misleading prompts are used to deceive the model; "jailbreak" attacks, aimed at extracting sensitive information; and "data poisoning" attacks, intended to alter the model's outputs.

Beyond the Hype of the LLMs – Why Are They So Popular?

In the last year, significant advancements in technology have been observed, particularly in the fields of large language models (LLM) and related artificial intelligence (AI) technologies. The widespread adoption of generative AI platforms is revolutionizing the way individuals and businesses gather and process information. This revolution is notably impacting sectors such as **industrial markets**, **enterprise analytics**, **healthcare**, **financial services**, **customer relations**, and **education**.

Current industry analyses forecast that the AI sector's value, which was at $11.3 billion in 2023, is projected to escalate to $51.8 billion by 2028. Given that ChatGPT has become the fastest-growing Internet application ever, it's evident that AI and LLMs will persist and evolve rapidly. To truly appreciate the influence of LLMs in the global market, it's vital to have an in-depth knowledge of the LLM sector and its role in reshaping various industries.

Grasping the overall market dynamics and the impact of AI technologies on key sectors is crucial before incorporating large language models (LLMs) into an enterprise's operational processes. With projections indicating a fourfold increase in the LLM market's value by 2029, the opportunities for innovation, advancement, and growth are substantial.

LLMs offer industries worldwide the ability to automate intellectual tasks, provide round-the-clock customer engagement, and extract vital insights from significant data and analytics. They support businesses in predicting market trends while minimizing the time, resources, and expenses involved.

These models are tailored with industry-specific data, enabling them to deliver precise and reliable outcomes in fields that demand intricate domain knowledge. LLMs, when configured appropriately, can process vast text datasets and complex structures. This capability provides organizations with more dependable insights, freeing up resources to concentrate on other areas and amplify business growth opportunities.

Common Benefits – The Real Reason Why LLMs Are So Popular

A key advantage that LLMs offer across various industries is the **enhancement of customer experiences**. By utilizing automated chatbots and interactive platforms, these models facilitate quicker and easier communication channels for consumers, ensuring their questions and requirements are efficiently met. Businesses benefit from increased customer satisfaction, loyalty, and retention by offering continuous, prompt response and support to their customers.

Large language models like GPT-4 have become popular for other key reasons as well:

- **Advanced Natural Language Understanding and Generation:** These models demonstrate a remarkable ability to understand and generate human-like text. This capability enables a wide range of applications, from writing assistance to conversational agents.

- **Versatility and Adaptability:** Large language models can be applied to various tasks without needing task-specific programming. They can perform tasks like translation, summarization, question answering, and even creative writing.

- **Ease of Integration:** These models can be integrated into existing software and services relatively easily, providing immediate enhancements to user experience and functionality.

- **Continual Learning and Improvement:** As more data becomes available and models are further trained, their performance and capabilities improve, making them increasingly valuable over time.

- **Business Efficiency:** They can automate and enhance tasks that previously required significant human effort, such as customer service, content creation, and data analysis, leading to cost savings and efficiency gains for businesses.

- **Accessibility of AI Technology:** With the availability of APIs and cloud-based services, these models are more accessible to a broader range of users and developers, fostering innovation and wider adoption.

- **Public Interest and Media Coverage:** The impressive capabilities of these models have garnered significant media attention, increasing public awareness and interest in AI.

- **Research and Development Investment:** Significant investment in AI research and development from both private and public sectors has driven rapid advancements in this field.

- **Personalization and User Engagement:** LLMs offer personalized responses, adapting to individual user styles and preferences. This enhances user engagement, making interactions more relatable and effective.

- **Support for Multiple Languages:** Many LLMs are trained on datasets encompassing various languages, making them versatile tools for global communication and translation services.

- **Enhanced Creativity and Idea Generation:** LLMs can assist in creative processes such as brainstorming, writing, and design, providing novel ideas and perspectives that can inspire human creativity.

- **Democratization of Knowledge and Expertise:** By providing access to information and expert-level guidance across various fields, LLMs democratize knowledge, making expert advice more accessible to a broader audience.

- **Educational Applications:** LLMs are increasingly used in educational settings for tutoring, language learning, and assisting students with learning disabilities, thereby enhancing educational experiences.

- **Advancements in Conversational AI:** The improved conversational abilities of LLMs have led to more natural and effective voice assistants and chatbots, revolutionizing customer service and personal assistants.

- **Data Analysis and Insights:** LLMs can process and analyze large volumes of text data, providing valuable insights and aiding in decision-making processes in business and research.

- **Accessibility for People with Disabilities:** LLMs offer tools that improve accessibility, such as generating summaries for long texts or providing voice-to-text services, thus aiding individuals with various disabilities.

- **Scalability:** LLMs can handle a large number of simultaneous interactions or requests, making them scalable solutions for businesses and organizations of all sizes.

- **Continual Evolution and Learning:** The field of LLMs is rapidly evolving, with frequent updates and improvements that continuously expand their capabilities and applications.

- **Collaboration and Community Building:** LLMs facilitate collaboration and community building by connecting people across language barriers and providing platforms for shared learning and interaction.

- **Impact on Art and Creativity:** Beyond text, some LLMs are venturing into artistic domains, aiding in music composition, visual arts, and creative writing, thereby influencing the artistic landscape.

- **Reduction in Human Error:** By automating routine tasks and providing accurate information retrieval, LLMs help reduce human error in various fields.

- **Cost-Effective Solutions:** For many businesses and users, LLMs offer cost-effective solutions by automating tasks that would otherwise require significant human labor and resources.

- **Enhanced Research Capabilities:** In academic and scientific research, LLMs aid in literature review, hypothesis generation, and even writing research papers, significantly enhancing research capabilities.

- **Environmental Monitoring and Sustainability:** Some LLMs are used in environmental monitoring, analyzing large datasets to aid in sustainability efforts and climate change research.

- **Entertainment and Gaming:** In the entertainment sector, LLMs contribute to game development, storytelling, and interactive media, enhancing user experience.

- **Public Policy and Governance:** LLMs assist in analyzing public opinion, drafting policy documents, and managing large-scale public communication, aiding in governance and policy-making.

- **Ethical and Philosophical Exploration:** The rise of LLMs has sparked important conversations about AI ethics, the future of work, and the philosophical implications of advanced AI.

- **Global Communication and Collaboration:** LLMs facilitate global communication and collaboration, breaking down language barriers and connecting people worldwide.

AI and large language models (LLMs) have revolutionized the way individuals and organizations interact with digital technology. These advancements propel innovation and automate processes across various industries, saving time and simultaneously altering the decision-making processes and customer communication strategies of professionals. They have reshaped specific industry domains, bolstering industrial advancement and the potential for innovation. As research and development continue to progress, **AI-driven models are on track to emulate human speech and interaction characteristics**.

Large Language Models for Business
Creation of Digital Content

Language models are adept at crafting a wide array of high-quality digital content, ranging from blog entries and articles to product descriptions and social media content, thereby conserving resources and time for businesses. They are also frequently employed in editing and correcting text. Serving as digital writing aids, these models offer instantaneous advice on grammar, spelling, stylistic enhancements, and alternative wording.

In addition, these models assist in the generation of creative content ideas and structure by evaluating existing materials, current trends, and audience preferences. This feature enables content creators to conceive novel and pertinent concepts that appeal to their target audience.

Furthermore, large language models are capable of condensing extensive texts into brief, succinct summaries swiftly, which is extremely useful for producing summarized versions of lengthy documents or executive summaries in a professional setting.

Enhancing Search Engine Optimization (SEO)

Language models are instrumental in refining SEO strategies, helping in several ways:

- Recommending appropriate primary and long-tail keywords to improve content's search engine visibility.

- Identifying frequently used search queries, enabling businesses to shape their content to better match user intent.

- Tailoring content for voice search queries, in line with the growing trend of voice-based searches.

- Enhancing meta descriptions and tags, critical for attracting clicks in search results.

- Structuring content to boost its discoverability and search engine rankings.

- Suggesting related terms and current topics, aiding businesses in creating content that engages and resonates with contemporary audiences.

- Ensuring websites are structured for easy indexing and crawling by search engine bots, thus improving visibility in search results.

- Performing SEO audits to assess various website components, such as loading speed, mobile compatibility, and URL structure, identifying potential enhancements.

- Incorporating the insights provided by LLMs into SEO strategies can lead to increased user engagement, longer duration spent on websites or apps, and overall improvement in online searchability and visibility.

Content Moderation

In the realm of automated content regulation, large language models (LLMs) are integral, adeptly spotting and sifting through user-generated material on various digital platforms. Their capabilities include recognizing problematic elements such as offensive language, hate speech, threats, misinformation, spam, and other forms of harmful or inappropriate content, thereby fostering a safer digital environment. These models can either mark such content for further inspection or autonomously eliminate it based on established moderation policies.

It's crucial to acknowledge that despite the substantial aid LLMs provide in content regulation, the necessity for human supervision remains, especially for complex or nuanced instances.

Emotion/Sentiment Analysis

LLMs are proficient in evaluating customer emotions through social media posts, reviews, and feedback. This process enables businesses to understand customer viewpoints and levels of satisfaction, an invaluable asset in monitoring social media and managing brand image.

These models classify emotional tone into various categories like positive, negative, or neutral and offer detailed emotion analysis by identifying varying intensities and subtle expression differences. This leads to a deeper comprehension of the emotions conveyed in texts.

Additionally, they are evolving in understanding context, capable of recognizing sarcasm, irony, and other complex language forms. While there has been notable progress, the area is still developing, with ongoing enhancements expected in the future.

Client Services

LLMs significantly contribute to customer service by improving and automating different facets of client interactions.

These include the following:

- **Chatbots** that handle queries, provide information, guide troubleshooting, and manage standard customer service tasks continuously

- **Voice assistants** for natural interaction through voice commands, performing tasks, retrieving information, and offering personalized aid

- **Virtual sales assistants** that interact with customers, answer queries about products, give recommendations, and navigate them through the buying process

Language Translation

LLMs have transformed language translation, offering effective and accurate translation capabilities. Their key feature is real-time translation of spoken or written material, invaluable in live conversations, international events, or instantaneous customer support.

Moreover, LLMs can specialize in specific domains or industries to heighten translation precision, including sector-specific terminology. Utilizing LLMs, businesses can communicate effectively with international clientele, bridge language gaps, and venture into new markets.

Virtual Teamwork

Incorporating LLMs in workplaces significantly boosts staff productivity and collaboration. They are essential in virtual teamwork and streamlining routine operations, capable of

- Generating summaries and transcripts of meetings
- Offering real-time translation for multilingual teams
- Assisting team members with disabilities, like visual or auditory impairments
- Documenting organizational and project processes
- Organizing company documentation
- Facilitating knowledge exchange

Recruitment and HR Support

LLMs are transforming recruitment and HR support, enhancing various aspects of hiring and aiding HR professionals. Their impact areas include the following:

- **Resume Analysis:** LLMs scrutinize resumes for relevant data such as skills, experience, and qualifications, aiding HR in efficiently shortlisting candidates.
- **Candidate Sourcing:** They assist in finding potential candidates from diverse platforms like job boards and social networks.
- **Profile Matching:** LLMs match job requirements with candidate profiles, pinpointing those who best fit the criteria.
- **Online Interviews:** They support virtual interviews by suggesting questions, evaluating responses, and summarizing discussions.
- **New Employee Integration:** LLMs guide new hires with information, training resources, and company policy orientation.

Sales Enhancement

LLMs significantly aid sales teams by providing insights and support throughout the sales cycle, including the following:

- **Lead Discovery:** Analyzing extensive data to pinpoint potential leads, understanding customer preferences for targeted engagement.

- **AI Chatbots:** Engaging site visitors, gathering information, offering insights to sales teams, and generating leads.

- **Customized Outreach:** Assisting in crafting personalized sales messages and recommendations for better conversion rates.

- **Analyzing Customer Feedback:** AI tools evaluate customer responses for a more tailored sales approach and stronger relationships.

Fraud Identification

LLMs excel in textual analysis, pattern recognition, and anomaly detection, making them effective in risk assessment and fraud prevention. Their real-time monitoring of data streams like financial transactions and client interactions allows quick identification of irregular or suspicious patterns, triggering immediate alerts for investigation.

Moreover, they assign risk scores to various transactions or activities based on a broad range of data, aiding in fraud likelihood assessment.

The future scope and capabilities of AI remain uncertain, yet its potential for innovation and progress seems boundless. The swift expansion of AI in business and industry sectors indicates that we are just beginning to uncover its full potential.

As AI functions evolve to be quicker and more efficient, sectors like healthcare, education, and financial services are set to flourish even more, providing reliable and trustworthy care and services to patients, students, and clients globally. LLMs, offering key support in data analysis and analytics, will lead to cost reductions as professionals redirect their focus and efforts. This era is marked by thrilling technological advancements, with both users and developers exploring the future direction of business and technology.

CHAPTER 2   WHAT ARE LARGE LANGUAGE MODELS?

Summary

This chapter delves into the concept of large language models (LLMs), exploring their development, capabilities, and underlying technologies. It also describes the architecture of LLMs, which includes various neural network layers such as recurrent layers, feedforward layers, embedding layers, and attention layers. These layers work together to process input text and generate output predictions.

The development of LLMs is attributed to advances in deep learning, substantial computational resources, and the abundance of training data. These models, typically pre-trained on extensive web corpora, can grasp intricate patterns, linguistic subtleties, and semantic relationships. Fine-tuning these models for specific tasks has led to state-of-the-art performance across various benchmarks.

In the following chapter, we'll get deeper into

- What is Python and its syntax design principles
- How Python is installed on Windows, Mac, and Linux
- Python variables and data types
- Python booleans and operators
- Conditionals and loops
- Python data structures
- Python functions
- OOP with Python
- Modules and file handling
- The powerful features of Python 3.11

CHAPTER 3

Python for LLMs

Python is a user-friendly and robust programming language, known for its high-level, efficient data structures and a straightforward approach to object-oriented programming. Its refined syntax and dynamic typing, combined with its capability to be interpreted, make Python a perfect choice for scripting and quick application development across various areas on numerous platforms.

In this chapter, we'll review Python's significance for large language models (LLMs). The chapter also delves into Python's syntax and semantics, noting its resemblance to languages like Perl, C, and Java while highlighting its unique features. But first, let's start with a quick primer.

Python at a Glance

Working with computers frequently leads to the realization that automating certain tasks can be highly beneficial. For instance, one might want to automate the process of editing multiple text files; other tasks could include creating a basic database, a unique GUI application, building a simple game, or dealing with data science, machine learning, and AI in your daily work.

Professional software developers often face the need to interact with various C/C++/Java libraries, but the standard cycle of writing, compiling, testing, and re-compiling can be inefficient. Writing test suites for such libraries and adding an extension language to a program are other common tasks where automation and simplicity are desired. In these scenarios, Python emerges as an ideal solution.

Python, on the other hand, is not only easier to use but is also available across various operating systems including Windows, Mac OS X, and Unix. Its simplicity doesn't take away from its capabilities as a real programming language, offering more structure and support for larger programs than what is possible with shell scripts or batch files.

Python's modularity allows for the reuse of code in other programs, enhancing efficiency. It comes with a vast array of **standard modules** that can serve as a foundation for your programs or as examples to learn Python programming. These modules cover a wide range of functionalities, including file I/O, system calls, sockets, and interfaces to GUI toolkits like Tk.

Being an interpreted language, Python **cuts down on development time**, as it doesn't require compilation and linking. Its interactive nature facilitates easy experimentation with language features, writing of temporary programs, or testing functions in a bottom-up approach to program development. Python also serves as a convenient desktop calculator.

Python's ability to enable **compact and readable programs** is one of its key strengths. Programs in Python are typically shorter than their equivalents in C, C++, or Java. This brevity is due to several factors: high-level data types allow for complex operations in single statements, indentation rather than brackets for statement grouping, and no need for declaring variables or arguments.

Python Syntax and Semantics

The rules and conventions governing the writing and interpretation of Python programs, both by computers and human readers, **constitute the syntax of the Python programming language**. Python shares several characteristics with languages like Perl, C, and Java, yet it also exhibits distinct differences. It is versatile in supporting various programming approaches such as structured, object-oriented, and functional programming. Additionally, Python is known for its **dynamic typing and automated memory management**.

Python is celebrated for its **straightforward and uniform syntax, guided by the philosophy that there should be a single, clear way to accomplish a task.** This language integrates native data types, control structures, first-class functions, and modules, enhancing code reuse and organization. Unlike many languages that rely heavily on punctuation, Python opts for **English keywords**, contributing to its clean and easy-to-understand appearance.

Python excels in error management with its comprehensive **exception handling mechanism and includes a debugger in its standard library, streamlining the debugging process**. The design of Python's syntax, emphasizing clarity and user-friendliness, has made it a favored choice for both novice and experienced programmers.

Interestingly, Python's name is derived from the BBC show "Monty Python's Flying Circus," and it has no association with snakes. In fact, incorporating references to Monty Python's sketches in documentation is not just permitted but actively encouraged.

Syntax Design Principles

Python is recognized for its **high readability as a programming language**. The simplicity and consistency of Python's syntax design are at its core, guided by the philosophy of "*There should be one – and preferably only one – obvious way to do it,*" a principle drawn from the Zen of Python.[1]

Zen of Python

- *Beautiful is better than ugly.*
- *Explicit is better than implicit.*
- *Simple is better than complex.*
- *Complex is better than complicated.*
- *Flat is better than nested.*
- *Sparse is better than dense.*
- *Readability counts.*
- *Special cases aren't special enough to break the rules.*
- *Although practicality beats purity.*
- *Errors should never pass silently.*
- *Unless explicitly silenced.*
- *In the face of ambiguity, refuse the temptation to guess.*
- *There should be one – and preferably only one – obvious way to do it.*
- *Although that way may not be obvious at first unless you're Dutch.*
- *Now is better than never.*

---

[1] "PEP 20 - The Zen of Python". Python Software Foundation. 2004-08-23

- *Although never is often better than right now.*
- *If the implementation is hard to explain, it's a bad idea.*
- *If the implementation is easy to explain, it may be a good idea.*
- *Namespaces are one honking great idea – let's do more of those!*

Python Identifiers

In Python, an identifier refers to a name given to entities like variables, functions, classes, modules, or other objects. It begins with a letter (A–Z or a–z) or an underscore (_), followed by any combination of letters, underscores, and digits (0–9). **Python identifiers cannot include special characters such as @, $, or %.**

Python is **case-sensitive**, meaning identifiers like "Hello" and "hello" are considered distinct.

Python follows specific naming conventions for identifiers:

- Class names in Python start with an uppercase letter, whereas all other identifiers begin with a lowercase letter.
- An identifier starting with a single underscore is considered private.
- An identifier beginning with two underscores is deemed strongly private.
- Identifiers that start and end with two underscores are reserved for special names defined by the language.

Table 3-1 lists 35 keywords or reserved words that are not available for use as identifiers.

Table 3-1. Python reserved words

and	as	assert	async	await	Break	class
continue	def	del	elif	else	Except	False
finally	for	from	global	if	Import	in
is	lambda	None	nonlocal	not	or	pass
raise	return	True	try	while	With	yield

CHAPTER 3   PYTHON FOR LLMS

Python Indentation

Python employs whitespace for defining blocks of control flow, adhering to the off-side rule. This characteristic of using indentation rather than punctuation or keywords to signify a block's extent is inherited from its forerunner, ABC.

In computer programming, a language is described as following the off-side rule syntax if the structure of its blocks is determined by indentation. This term was introduced by **Peter Landin** and is thought to be a play on words relating to the offside rule in soccer. This concept differs from free-form languages, especially those using curly brackets, where indentation doesn't carry computational significance and is purely a matter of coding style and format. Languages that apply the **off-side rule** are often characterized by their significant indentation.

Contrastingly, in "free-format" languages that evolve from the ALGOL block structure, code blocks are demarcated using braces ({ }) or specific keywords. Typically, in these languages, coders indent the contents of a block as a conventional practice to visually distinguish it from the adjacent code.

Example:

```
def greetings():
    print("hello world")

for i in range(10)
    print(i)
```

In Python, adhering to specific rules is crucial when it comes to code indentation:

- Avoid splitting indentation across multiple lines using the backslash ("\") character.

- You must not indent the first line of Python code; doing so will result in an Indentation Error.

- Consistency is key when choosing between tabs and spaces for indentation. Stick with your preferred spacing method throughout your code. Mixing tabs and spaces can lead to incorrect indentation.

- It's generally recommended to use either one tab or four whitespaces for the initial level of indentation and increment by an additional four whitespaces (or one more tab) for deeper indentation levels.

105

Python Multiline Statements

In Python, statements usually terminate with a newline character. However, Python offers the flexibility of using the line continuation character (\) to indicate that a line should be continued.

Example:

```
total = one + \
        two + \
        three
```

Statements enclosed within square brackets [], curly braces {}, or parentheses () do not require the line continuation character. For instance, the following statement is valid in Python:

Example:

```
days = ['Monday', 'Tuesday', 'Wednesday',
        'Thursday', 'Friday']
```

Quotations in Python

In Python, string literals can be represented using single ('), double ("), or triple (''' or """) quotes, provided that the same type of quote is used to both start and end the string.

Triple quotes, on the other hand, are employed to encompass strings that span multiple lines. For instance, the following examples are all valid in Python:

Example:

```
name = 'Alice'
print(f"My name is {name}.")

quote = "To be or not to be, that is the question."
print(f"Shakespeare once said, '{quote}'")

story = """Once upon a time,
there was a brave knight
who embarked on a grand adventure."""
```

```
print(story)
```

Output:

```
My name is Alice.
Shakespeare once said, 'To be or not to be, that is the question.'
Once upon a time,
there was a brave knight
who embarked on a grand adventure.
```

Comments in Python

In Python, comments serve as programmer-readable explanations or annotations within the source code. They are included to enhance the code's comprehensibility for human readers and are disregarded by the Python interpreter.

Python, like many contemporary programming languages, supports both single-line (end-of-line) and multiline (block) comments. The structure of Python comments closely resembles those found in languages such as PHP, BASH, and Perl.

To create a comment in Python, you use the hash sign (#). Any hash sign that is not within a string literal marks the beginning of a comment. Everything following the # symbol up to the end of the physical line is considered part of the comment, and Python's interpreter disregards it.

```
# First comment
print ("Hello, World!") # Second comment
```

You can type a comment on the same line after a statement or expression:

```
Name = "John" # comment on the same line
```

You can comment multiple lines as follows:

```
# Comment 1
# Comment 2
# Comment 3
```

The Python interpreter will also ignore the triple-quoted string that follows, allowing it to serve as a multiline comment:

```
'''
This is my first multiline
comment.
'''
```

Utilizing Blank Lines in Python Code

A blank line in Python code is one that consists solely of whitespace or possibly contains a comment, and Python completely disregards it. When working in an interactive interpreter session, you must input an empty physical line to conclude a multiline statement.

Combining Multiple Statements in a Single Line

You can use the semicolon (;) to include multiple statements on a single line, provided that none of these statements initiates a new code block. Here's an example snippet demonstrating the use of the semicolon:

Example:

```
x = 5; y = 10; z = x + y; print(z)
```

How to Install Python and Your First Python Program

Installing Python may differ depending on the operating system you are using. Therefore, installation instructions for the following operating systems have been provided:

- Windows
- macOS
- Linux

CHAPTER 3   PYTHON FOR LLMS

> **Note** Python Installation on Different Operating Systems
>
> *Please note that the installation steps can vary depending on your specific operating system and its version. To install Python on your particular platform, follow the instructions provided here for your respective operating system.*

Installing Python on Windows

This section provides a step-by-step guide for installing Python on a Windows operating system. Follow these instructions to successfully set up Python on your Windows computer:

Step 1: Download the Python Installer

First, visit the official Python website to procure the latest Python 3.x version tailored for Windows. The website will automatically detect your operating system and offer the suitable installer (32-bit or 64-bit) for your system.

Step 2: Launch the Installer

Locate the downloaded installer file, typically found in your Downloads folder, and double-click on it to initiate the installation process. You might encounter a User Account Control (UAC) prompt asking for permission to proceed. Click "Yes" to continue.

Step 3: Customize the Installation (Optional)

Upon reaching the installer's welcome screen, you'll encounter two options: "Install Now" and "Customize installation" (Figure 3-1).

If you prefer a default installation, with standard settings, simply click "Install Now."

CHAPTER 3 PYTHON FOR LLMS

Figure 3-1. *Step 1 of the installation process of Python*

For a customized installation where you can modify parameters such as the installation directory or select specific components, click "Customize installation" (Figure 3-2).

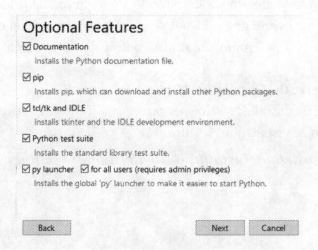

Figure 3-2. *Optional features during installation*

This choice presents the following options:

- **Documentation:** This includes Python's documentation file alongside the installation.

CHAPTER 3   PYTHON FOR LLMS

- **pip:** This option installs pip, enabling you to install additional Python packages as needed.

- **Tkinter and IDLE** (tcl/tk and IDLE): This option installs tkinter and IDLE.

- **Python test suite:** Selecting this option installs the standard library test suite, helpful for testing your code.

- **py launcher (for all users):** These options allow you to launch Python from the command line.

Once you've made your selections, click "Next." A new dialog box with advanced options will appear, offering further choices as shown in Figure 3-3.

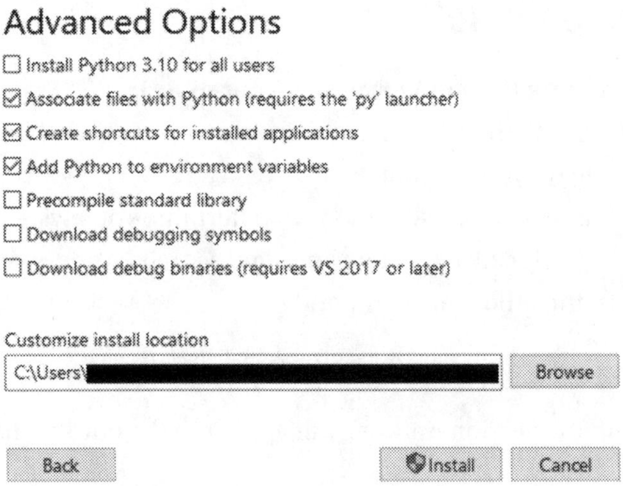

Figure 3-3. Advanced options during installation

- Install Python 3.11 for all users.

- Associate files with Python (requires the "py" launcher).

- Create shortcuts for installed applications.

- Add Python to environment variables.

- Precompile standard library.

- Download debugging symbols.

- Download debug binaries (requires VS 2017 or later).

111

Ensure that the chosen installation directory is correct, then proceed to the next step.

Step 4: Commence Python Installation

After configuring your preferred installation settings, click "Install" to initiate the installation procedure. The installer will copy essential files to your computer and configure Python. This process may take a few minutes.

Step 5: Confirm the Installation

Once the installation concludes, confirm that Python has been successfully installed by opening the Command Prompt (search for "cmd" in the Start menu) and entering the following command:

```
python --version
```

Install Python on macOS

Step 1: Verify the Existing Python Version on Your macOS

Before proceeding with the Python installation on macOS, it's wise to confirm the current Python version on your system. macOS typically includes an older Python version (Python 2.x) as a default installation. To determine your system's Python version, access the Terminal app (locate it through Spotlight search or under Applications ➤ Utilities). Then, input the following command:

```
python --version
```

Press "Enter," and the version number will appear in the output, displayed as follows:

```
Python 2.7.x
```

If Python 3.x is already installed on your system, you can ascertain its version by executing the subsequent command:

```
python3 --version
```

Step 2: Visit the Python Site and Download the macOS installer

Navigate to the download page, where you'll find the macOS installer package in the form of a .pkg file, corresponding to the latest Python release. Proceed to download the installer to your computer.

Locate the installer file you've downloaded, typically located in your Downloads folder, and double-click it to initiate the installation procedure. Subsequently, follow the on-screen instructions as prompted.

Step 3: Verify Python and IDLE Are Installed Correctly

After the installation process is finished, a folder will open on your desktop. Inside this folder, click on the "IDLE" application. IDLE is the stand-alone development environment that comes bundled with Python. Upon launching IDLE, the Python shell should open automatically.

To verify that it's functioning correctly, you can test it by entering a print command as shown here:

```
print("Test message")
```

Press "Enter," and you should observe the text "Test message" appearing on the subsequent line within the IDLE environment. You can also confirm the installation through the Terminal application. Open Terminal and input the following com mand:

```
python3 --version
```

Press "Enter," and you should see the version of Python that you recently installed displayed in the output. This serves as confirmation that Python has been successfully installed on your Mac.

Installing Python on Linux – Ubuntu/Debian and Fedora

Step 1: Checking for Pre-installed Python

To determine whether Python is already installed on your system, open a terminal window and execute the following command:

```
python --version
```

Step 2: Installing Python via Package Manager

The most straightforward method to install Python on a Linux system is by utilizing the Package Manager that corresponds to your distribution. Here are a couple of commonly used commands for popular distributions:

For Ubuntu/Debian, use

```
sudo apt-get install python3
```

For Fedora, execute

```
sudo dnf install python3
```

These commands will help you effortlessly install Python on your Linux distribution using the respective Package Manager.

CHAPTER 3   PYTHON FOR LLMS

Your First Python Program

You have the choice of using a simple code editor known as IDLE or an integrated development environment (IDE), which is part of the Python installation package. There are various ways to run Python interactively. You can opt for the system's Console, the Python IDLE (Figure 3-4), or the Python Shell, all of which are included in the standard Python installation available at www.python.org.

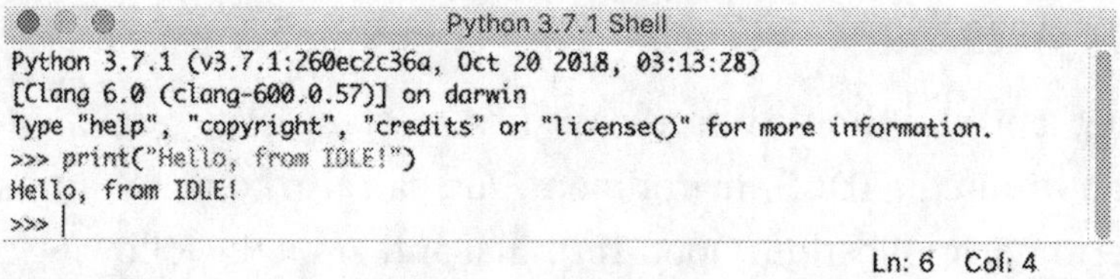

Figure 3-4.   First Python program in Python Shell

To **write your first program**, open it and type the following line of code, then hit Enter:

```
print("Hello, from IDLE!")
```

You can activate the Python interpreter by opening your operating system's terminal application and typing "python3". This command will launch the Python interpreter in your terminal, allowing you to interact with Python directly from the command line.

Python offers versatility in its usage, allowing you to utilize it in two primary modes: **interactive mode and scripting mode**. The Python program you've installed inherently functions as an interpreter. An interpreter processes text commands and executes them as you input them, making it particularly convenient for experimentation and rapid testing.

Variables and Data Types, Numbers, Strings, and Casting

Variables serve as labels for data that we want to store and manipulate within our programs. Let's consider an example: suppose your program must retain a user's name. To achieve this, we can create a variable named "user_name" and initialize it using the following statement:

```
user_name = "John"
```

Once you've defined the `"user_name"` variable, **your program will allocate a specific portion of your computer's memory to hold this data**. You can subsequently access and modify this data by referring to it with its age, `"user_age"`. When introducing a new variable, you must provide it with an initial value. In this instance, we've assigned it the value "John". It's important to note that **you can always alter this value within your program later on**.

Moreover, you can define multiple variables in a single step. To do this, simply write

```
use_rage, user_name = 30, 'Peter'
```

Naming a Variable

In Python, variable names must adhere to certain rules and conventions:

1. **Variable names can contain letters** (both lowercase and uppercase), numbers, and underscores (_). However, they must begin with a letter or underscore, and the first character cannot be a number. For example, valid variable names are "userName," "user_name," or "userName2," **but "2userName" is not allowed**.

2. **Certain words are reserved in Python** because they already have predefined meanings within the language. These reserved words include terms like "print," "input," "if," "while," and all others previously mentioned as reserved keywords. You cannot use these reserved words as variable names. We'll explore these reserved words in greater detail in subsequent chapters.

3. **Variable names in Python are case-sensitive**. This means that "username" and "userName" are considered distinct variables.

When it comes to naming variables in Python, there are two common conventions to follow:

- **Camel Case Notation:** This practice involves writing compound words with mixed casing, where each word begins with a capital letter except the first one. For example, `"thisIsAVariableName"`. This convention will be used throughout the remainder of this book.

- **Underscore Separation:** An alternative convention is to use underscores (_) to separate words in variable names. For those who prefer this style, you can name your variables like "this_is_a_variable_name".

- **Pascal Case:** Identical to Camel Case, except the first word is also capitalized. Example: `NumberOfUsers`.

Ultimately, adhering to these rules and conventions ensures that your variable names are both valid and readable within the Python programming language.

Using underscores to separate words in variable names, also known as snake_case, is more widely used as the naming convention for Python variables. This convention is recommended by the official Python style guide, PEP 8, which is widely followed by the Python community. Snake_case is considered more Pythonic and is the preferred style for variable names in most Python projects and libraries.

Data Types

Numbers in Python

There are three numeric types in Python:

- int
- float
- complex

```
x = 10   # int
y = 2.8  # float
z = 3 + 2j  # complex
```

Integers

Integers represent whole numbers, whether positive or negative, without any decimal component. For instance, in the field of neuroscience, researchers might use integers to tally the count of active neurons at a specific moment.

Floats

Floats, also known as floating-point real values, encompass real numbers that can include decimal points. For instance, recording a participant's reaction time in milliseconds for a specific task would involve using floating-point numbers.

Complex

In Python, complex numbers are a built-in data type used to represent numbers in the form "a + bj," where "a" and "b" are real numbers and "j" represents the imaginary unit, equal to the square root of -1. Complex numbers are often used in mathematical and scientific computations where real and imaginary parts need to be handled simultaneously.

You can perform various operations on complex numbers in Python, including addition, subtraction, multiplication, division, and more. Python's standard library provides functions and methods for working with complex numbers.

Note Numeric Variables

They are automatically generated when you assign values to them, making it straightforward in regular coding. Therefore, you can assign values to variables without concerning yourself with specifying their data type.

Strings

A string is a collection of characters, which can include letters, numbers, symbols, or spaces, enclosed within either "single" or "double" quotation marks in Python. It's important to note that strings in Python are **immutable**, which means once a string is created, you cannot modify its contents directly; instead, you can only overwrite it by redefining the variable.

```
>>> a = "Hello World!"
>>> print(a)
>>> Hello World!
```

You can display strings on the screen using the print function. In Python, strings are sequences of bytes, and you can access a specific character within a string using indexing or in Python called "slicing," as demonstrated in the code example:

```
>>> a = "Hello World!"
>>> print(a[0:2])
>>> He
```

Note that the last character, when slicing a string, is exclusive!

There are many built-in functions form manipulating strings in Python; part of them is as follows:

- split(): Which splits the string by a character or empty space and makes it a list
- replace(): To replace one character in the string with another one
- upper(): To make all characters of the string uppercase
- lower(): To make all characters of the string lowercase
- len(): To get the length of the string

String Delimiters and Characteristics

In Python, you can define string literals using either single or double quotes. Any characters enclosed between the opening delimiter and the corresponding closing delimiter are considered part of the string. There's no strict limit on the length of a string in Python. It can contain as many characters as your machine's memory resources permit. Furthermore, a string can even be empty.

How do you handle situations where you need to include a quotation mark as part of the string itself? As you can observe, the straightforward approach encounters issues. In the given example, the string begins with a single quote, leading Python to interpret the following single quote within the parentheses as the closing delimiter, unintentionally treating it as part of the string. Consequently, the final single quote becomes an extraneous character, resulting in a syntax error:

```
>>> print('This is a single quote (') character.')
SyntaxError: invalid syntax
```

To include either type of quotation mark within the string, a simple and effective method is to enclose the string with the opposite type of quotation marks. ***If you wish to incorporate a single quote, surround the string with double quotes, and conversely, if you need to include a double quote, enclose the string within single quotes.*** This approach ensures that the desired quotation mark is correctly interpreted as part of the string.

```
>>> print("This contains a single quote (') character.")
This string contains a single quote (') character.

>>> print('This contains a double quote (") character.')
This string contains a double quote (") character.
```

Handling Special Characters in Strings

In certain situations, you may need Python to interpret specific characters within a string differently.

This can happen in two ways:

- **Suppressing Special Interpretation:** You might want to prevent certain characters from having their usual special meaning within a string.

- **Applying Special Interpretation:** Alternatively, you may wish to give special meaning to characters that are typically treated literally.

To achieve this, you can use the backslash (\) character. When a backslash appears in a string, it signals that one or more characters immediately following it should be treated in a special manner. This mechanism is known as an "escape sequence" because the backslash causes the subsequent character sequence to deviate from its standard interpretation.

```
>>> print('String contains a single quote (') character.')
SyntaxError: invalid syntax

>>> print('String contains a single quote (\') character.')
This string contains a single quote (') character.
```

RAW Strings

To denote a raw string literal, you use the prefix "r" or "R," indicating that escape sequences within the string will remain unaltered. This means the backslash character will not be interpreted as an escape character:

Raw String:

```
print(r'foo\nbar')
foo\nbar
```

Raw String:

```
print(R'foo\\bar')
foo\\bar
```

In a raw string, backslashes are not treated as escape characters, preserving their literal representation.

Triple-Quoted Strings in Python

Python offers an alternative method for defining strings known as triple-quoted strings. These strings are enclosed by three consecutive single quotes (''') or three consecutive double quotes ("""). While escape sequences remain functional within triple-quoted strings, you can include single quotes, double quotes, and even newlines without needing to escape them. This feature simplifies the creation of strings containing both single and double quotes:

For instance:

```
print('''This string has a single (') and a double (") quote.''')
This string has a single (') and a double (") quote.
```

In triple-quoted strings, the inclusion of single and double quotes does not necessitate escape characters, making it a convenient choice for such scenarios.

Booleans and Operators

Booleans

The Python Boolean type stands as one of the integral built-in data types within the Python programming language. Its primary purpose is to convey the truth value of a given expression, aiding in logical evaluations. For instance, the expression "1 <= 2" evaluates to True, while "0 == 1" evaluates to False. ***A comprehensive understanding of how Python Boolean values operate is crucial for effective Python programming.***

Python's Boolean type encompasses just two possible values:

- True
- False

No other value within Python assumes the bool type. You can ascertain the type of True and False by employing the built-in "type()" function:

```
>>> type(False)
<class 'bool'>
>>> type(True)
<class 'bool'>
```

The output <class 'bool'> indicates the variable is a Boolean data type. It's worth noting that the "bool" type is an intrinsic component of Python, eliminating the need for importing external libraries. However, the name "bool" itself is not a reserved keyword in the Python language.

Note True and False in Python

The keywords "True" and "False" in Python must always begin with an uppercase letter. Attempting to use a lowercase "true" or "false" will result in an error.

```
>>> x = true
Traceback (most recent call last):
  File "<input>", line 1, in <module>
NameError: name 'true' is not defined
```

Converting Integers and Floats into Booleans

In Python, you can transform integers and floating-point numbers into Boolean values by utilizing the built-in `bool()` function. When an integer, float, or complex number is assigned a value of zero, it yields a boolean outcome of False. Conversely, if the number is set to any other nonzero value, whether positive or negative, it evaluates to True.

For example:

```
zero_int = 0
bool(zero_int)
# Output: False

pos_int = 1
bool(pos_int)
# Output: True

neg_flt = -5.1
bool(neg_flt)
# Output: True
```

Boolean Operators

Boolean arithmetic revolves around the manipulation and combination of true and false logic values. In Python, boolean values are represented as either True or False, and you can perform operations on them using boolean operators.

The commonly used boolean operators in Python are as follows:

- or
- and
- not
- == (equivalent)
- != (not equivalent)

In the following code segment, two variables, A and B, are assigned boolean values of True and False, respectively. Subsequently, these boolean values are subjected to various operations using boolean operators:

```
A = True
B = False

A or B      # Result: True
A and B     # Result: False
not A       # Result: False
not B       # Result: True
A == B      # Result: False
A != B      # Result: True
```

Furthermore, you can create complex boolean expressions by **combining boolean operators and using parentheses for precedence**:

```
C = False
A or (C and B)    # Result: True
(A and B) or C    # Result: False
```

Table 3-2 represents a summary of boolean arithmetic and boolean operators.

Table 3-2. Boolean arithmetic

A	B	Not A	Not B	A == B	A != B	A or B	A and B
T	F	F	T	F	T	T	F
F	T	T	F	F	T	T	F
T	T	F	F	T	F	T	T
F	F	T	T	T	F	F	F

Python Operators

Operators are specialized symbols employed to execute operations on both values and variables. They encompass a set of unique symbols dedicated to carrying out arithmetic and logical computations. The item upon which the operator acts is referred to as the operand.

CHAPTER 3   PYTHON FOR LLMS

Arithmetic Operators

In Table 3-3, Python arithmetic operators serve the purpose of executing mathematical operations, including addition, subtraction, multiplication, and division.

Table 3-3. *Python arithmetic operators*

Operator	Description	Syntax
+	Addition: adds two operands	x + y
–	Subtraction: subtracts two operands	x – y
*	Multiplication: multiplies two operands	x * y
/	Division (float): divides the first operand by the second	x / y
//	Division (floor): divides the first operand by the second	x // y
%	Modulus: returns the remainder when the first operand is divided by the second	x % y
**	Power (Exponent): Returns first raised to power second	x ** y

Example:

```
>>> a = 3
>>> b = 2
>>> print(a + b)
5
>>> print(a - b)
1
>>> print(a * b)
6
>>> print(a % b)
1
>>> print(a ** b)
9
```

There are two types of division operators:

- Float division
- Floor division

CHAPTER 3 PYTHON FOR LLMS

Float Division Example:

```
>>> print(5/5)
>>> print(10/2)
>>> print(-10/2)
>>> print(20.0/2)
```

Output:

```
1.0
5.0
-5.0
10.0
```

Floor Division Example:

```
>>> print(10//3)
>>> print(-5//2)
>>> print(5.0//2)
>>> print(-5.0//2)
```

Output:

```
3
-3
2.0
-3.0
```

Comparison Operators

Table 3-4 shows a comparison using relational operators involving evaluating values and produces a result of either True or False, depending on whether the condition is met or not.

Table 3-4. Relational operators comparison

Operator	Description	Syntax
>	Greater than: True if the left operand is greater than the right	x > y
<	Less than: True if the left operand is less than the right	x < y
==	Equal to: True if both operands are equal	x == y
!=	Not equal to: True if operands are not equal	x != y
>=	Greater than or equal to True if the left operand is greater than or equal to the right	x >= y
<=	Less than or equal to True if the left operand is less than or equal to the right	x <= y

Example:

```
>>> a = 5
>>> b = 3
>>> print(a > b)
True
>>> print(a < b)
False
>>> print(a == b)
False
>>> print(a != b)
True
>>> print(a >= b)
True
>>> print(a <= b)
False
```

Logical Operators

Within the realm of Python, logical operators come into play when handling conditional statements, which typically revolve around either True or False outcomes, presented in Table 3-5. These operators are responsible for executing logical AND, logical OR, and logical NOT operations.

Table 3-5. Python logical operators

Operator	Description	Syntax	Example
And	Returns True if both the operands are true	x and y	x>5 and x>7
Or	Returns True if either of the operands is true	x or y	x<7 or x>21
Not	Returns True if the operand is false	not x	not(x>11 and x> 21)

Example:

```
>>> a = True
>>> b = False
>>> print(a and b)
False
>>> print(a or b)
True
>>> print(not a)
False
```

Bitwise Operators

Python's bitwise operators, presented in Table 3-6, function at the level of individual bits and execute operations that involve the manipulation of bits themselves. They find their utility in working with binary numbers.

Table 3-6. Bitwise operators

Operator	Description	Syntax
&	Bitwise AND	x & y
|	Bitwise OR	x | y
~	Bitwise NOT	~x
^	Bitwise XOR	x ^ y
>>	Bitwise right shift	x>>
<<	Bitwise left shift	x<<

Example:

```
>>> a = 10
>>> b = 4
>>> print(a & b)
0
>>> print(a | b)
14
>>> print(~a)
-11
>>> print(a ^ b)
14
>>> print(a >> 2)
2
>>> print(a << 2)
40
```

Assignment Operators

Assignment operators in Python, represented in Table 3-7, serve the purpose of assigning values to variables.

Table 3-7. *Assignment operators*

Operator	Description	Syntax	
=	Assign the value of the right side of the expression to the left-side operand	x = y + z	
+=	Add AND: Add right-side operand with left-side operand and then assign to left operand	a+=b	a=a+b
-=	Subtract AND: Subtract right operand from left operand and then assign to left operand	a-=b	a=a-b
*=	Multiply AND: Multiply right operand with left operand and then assign to left operand	a*=b	a=a*b
/=	Divide AND: Divide left operand with right operand and then assign to left operand	a/=b	a=a/b
%=	Modulus AND: Take modulus using left and right operands and assign the result to left operand	a%=b	a=a%b
//=	Divide(floor) AND: Divide left operand with right operand and then assign the value(floor) to left operand	a//=b	a=a//b
**=	Exponent AND: Calculate exponent(raise power) value using operands and assign value to left operand	a**=b	a=a**b
&=	Perform Bitwise AND on operands and assign value to left operand	a&=b	a=a&b
|=	Perform Bitwise OR on operands and assign value to left operand	a|=b	a=a|b
^=	Perform Bitwise xOR on operands and assign value to left operand	a^=b	a=a^b
>>=	Perform Bitwise right shift on operands and assign value to left operand	a>>=b a=a>>b	
<<=	Perform Bitwise left shift on operands and assign value to left operand	a <<= b a= a << b	

Example:

```
>>> a = 10
>>> b = a
```

```
>>> print(b)
10
>>> b += a
>>> print(b)
20
>>> b -= a
>>> print(b)
10
>>> b *= a
>>> print(b)
100
>>> b <<= a
>>> print(b)
102400
```

Identity Operators

In Python, "is" and "is not" are identity operators employed to verify whether two values occupy the same memory location as shown in Table 3-8. It's important to note that equality between two variables does not necessarily imply their identity.

Table 3-8. Identity operators

Operator	Description
Is	True if the operands are identical
is not	True if the operands are not identical

Example:

```
>>> a = 10
>>> b = 20
>>> c = a
>>> print(a is not b)
True
>>> print(a is c)
True
```

CHAPTER 3   PYTHON FOR LLMS

Membership Operators

Within Python, the "in" and "not in" operators are categorized as membership operators (Table 3-9), and their primary function is to assess whether a particular value or variable exists within a given sequence.

Table 3-9. *Membership operators*

Operator	Description
In	True if value is found in the sequence
not in	True if value is not found in the sequence

Example:

```
>>> x = 21
>>> y = 10
>>> list = [10, 20, 30, 40, 50]
>>> if (x not in list):
...     print("x is NOT present in given list")
>>> else:
...     print("x is present in given list")
>>> x is NOT present in given list
>>> if (y in list):
...     print("y is present in given list")
>>> else:
...     print("y is NOT present in given list")
>>> y is present in given list
```

Ternary Operator

A Ternary operator, also referred to as conditional expressions, is an operator designed to assess a condition as either true or false. It was introduced to Python starting from version 2.5. It is a concise way to evaluate a condition in a single line, thus replacing the need for a multiline if-else statement and resulting in more compact code.

Syntax: `[on_true] if [expression] else [on_false]`

Example:

```
>>> a, b = 10, 20
>>> min = a if a < b else b
>>> print(min)
>>> 10
```

Conditionals and Loops

Conditionals and loops are fundamental constructs in programming that enable developers to create dynamic, efficient, and responsive applications. Conditionals, also known as decision-making statements, allow a program to execute different actions based on specific criteria. By evaluating boolean expressions, conditionals such as "if", "else if", and "else" statements determine the flow of execution, making programs adaptable to varying inputs and conditions.

Loops, on the other hand, are iterative control structures that repeat a block of code as long as a given condition is true. Common types of loops include "for" and "while" loops. These constructs are crucial for tasks that require repetitive actions, such as processing elements in an array, handling user input until a valid response is received, or performing complex calculations iteratively. By leveraging conditionals and loops, programmers can design robust and flexible code, capable of handling a wide range of scenarios and improving overall program efficiency.

Conditionals

In the examples up to this point, you've accumulated a substantial amount of Python programming knowledge, focusing on code that executes in a linear sequence. This means each command is processed one after the other, in the precise sequence they are laid out.

However, real-world scenarios often demand more flexibility. Programs may need to bypass certain instructions, repeatedly execute a block of statements, or choose from among different sets of instructions depending on various conditions.

This is where the concept of **control structures** becomes crucial. Control structures are instrumental in guiding the flow of execution within a program, also known as its **control flow**.

Within the context of Python, the "if" statement is the fundamental mechanism for making decisions. It enables the conditional execution of a single statement or a collection of statements, contingent on the evaluation of an expression.

Example:

```
if <expr>:
    <statement>
```

In this code snippet:

- "`<expr>`" represents an expression that Python evaluates to determine its truth value, a concept elaborated upon in the discussion about logical operators within the Python Operators and Expressions tutorial.

- "`<statement>`" refers to any valid line of Python code that follows an indentation rule.

- When "<expr>" yields a truthy outcome (meaning Python interprets it as true), then "<statement>" is carried out. Conversely, if "<expr>" results in a falsy value (interpreted as false), then "<statement>" is bypassed and remains unexecuted.

- It's important to note the necessity of the **colon (":") after "<expr>"**. Unlike some languages that mandate the encapsulation of "<expr>" within parentheses, Python has no such requirement.

Example:

```
>>> x = 0
>>> y = 5
>>> if x < y:    # Truthy
...     print('yes')
yes
```

Grouping Statements

As mentioned earlier in this chapter, Python employs a programming principle known as the off-side rule. This rule is characterized using indentation to demarcate blocks of code. Python belongs to a relatively exclusive group of languages that implement the off-side rule.

CHAPTER 3   PYTHON FOR LLMS

The role of indentation is not merely stylistic but **carries functional significance in Python code**. The reason behind this is now clear: indentation serves to delineate compound statements or blocks within the code. **Therefore, in Python, lines of code that share the same level of indentation are treated as part of the same block.**

Consequently, the structure of a compound "if" statement in Python is defined through indentation:

```
1|  if <expr>:
2|      <statement>
3|      <statement>
4|      <statement>
        ...
5|  <following_statement>
```

In this case, statements that share the same indentation level (from lines 2 to 4) are grouped together as a single block. If <expr> evaluates to true, the entire block is executed; if <expr> is false, the block is bypassed. After processing this block, the program continues with <following_statement>.

Nested Blocks

Blocks within Python can be nested to any level of depth, where each additional indentation level signifies the beginning of a new block and each decrease in indentation marks the end of the current block. This hierarchical arrangement of code blocks creates a structure that is simple, uniform, and easy to understand.

Example:

```python
age = 20
has_license = True

if age >= 18:
    if has_license:
        print("You are eligible to drive.")

    print("You are not old enough to drive.")
```

Else and Elif Clauses

You've learned to employ an "if" statement for executing either a solitary instruction or a group of instructions based on a condition. Now, let's explore additional capabilities. At times, you might need to assess a situation and follow one course of action if the condition holds true, while also defining a different course should the condition prove false. This scenario is effectively managed by incorporating an "else" clause.

```
if <expr>:
    <statement(s)>
else:
    <statement(s)>
```

When <expr> evaluates to true, **the program executes the initial set of statements and bypasses the second set**. Conversely, if <expr> is false, the program skips the first set and **proceeds with the execution of the second set**. Following the completion of these conditional branches, the program's flow continues beyond the second set of statements.

Example:

```
temperature = 30

if temperature > 25:
    print("It's a hot day.")
else:
    print("It's not a hot day.")
```

Python offers a way to navigate through multiple conditional paths by employing one or more "**elif**" clauses, which stands for "*else if*." The language sequentially assesses each condition ("<expr>") and executes the block of code associated with the first condition that evaluates to true. If all conditions prove false and an "else" clause is present, then the code block under the "else" clause will be executed.

Example:

```
score = 75  # Assume this is the score out of 100

if score >= 90:
    print("Grade: A")
```

```
elif score >= 80:
    print("Grade: B")
elif score >= 70:
    print("Grade: C")
elif score >= 60:
    print("Grade: D")
else:
    print("Grade: F")
```

One-Line if Statements

Conventionally, "<expr>" in an "if" statement is written on one line, with the "<statement>" indented beneath it on the next line. However, it's also acceptable to construct the entire "if" statement on a single line, achieving the same functionality.

Syntax:

```
if <expr>: <statement>
```

It's possible to include multiple "<statement>"s on a single line by separating them with semicolons:

Syntax:

```
if <expr>: <statement_1>; <statement_2>; ...; <statement_n>
```

Python Loops (For and While)

A loop in programming is a fundamental concept that allows a block of code to be executed repeatedly based on a condition or until a certain condition is met. ***Loops are used to automate repetitive tasks, making it possible to perform operations on collections of data, generate sequences, or wait for certain events*** without the need for writing the same lines of code multiple times.

There are several types of loops, but the most common ones include the following:

- **For Loop**: Iterates over a sequence (such as a list, tuple, dictionary, set, or string) and executes a block of code for each item in the sequence. It's often used when the number of iterations is known or finite.

- **While Loop**: Continues to execute a block of code as long as a specified condition remains true. It's used when the number of iterations is not known before the loop starts and depends on when a certain condition changes.

Loops can include various control statements to modify their execution flow, such as "`break`" to exit the loop prematurely, "`continue`" to skip the current iteration and proceed to the next one, or "`else`" to execute a block of code once the loop condition is no longer true (applies to Python).

While Loop in Python

In Python, a "while" loop repeatedly executes a set of statements as long as a specified condition remains true. Once the condition evaluates to false, the program resumes execution with the line immediately following the loop.

Syntax:

```
while expression:
    statement(s)
```

Like with the if conditionals, Python groups statements into a single block of code based on the uniform indentation level following a programming construct. The same number of spaces used to indent statements determines their inclusion in the same code block.

Example:

```
counter = 0  # Initialize counter

while counter < 5:
    print("Counter is", counter)
    counter += 1  # Increment counter

print("Loop finished")
```

Output:

```
Counter is 0
Counter is 1
Counter is 2
```

```
Counter is 3
Counter is 4
Loop finished
```

Else Statement with while Loop

In Python, else statement could be used along with the while loops. Here is an example:

```
count = 0   # Initialize count

while count < 3:
    print("Count is", count)
    count += 1  # Increment count
else:
    print("Count is no longer less than 3")
```

In this example, the while loop executes as long as count is less than 3. It prints the current value of count and then increments count by 1 each time through the loop. When count reaches 3, the loop condition count < 3 becomes false, causing the loop to exit. At this point, the else block is executed, printing "Count is no longer less than 3". **The else part of a while loop runs when the loop condition becomes false naturally, meaning it wasn't exited through a break statement.**

Creating an Infinite Loop with Python while Loop

To repeatedly execute a code block an indefinite number of times, Python "while" loop can be employed for such a purpose. This approach involves utilizing a "while" loop with the condition "(count == 0)". The loop is designed to continue executing as long as the value of "count" remains 0. Given that "count" is initially set to 0, this leads to a loop that will run endlessly, as its condition perpetually holds true.

Caution   Employing such an endless loop is generally advised against because it creates a loop with no natural conclusion, where the condition remains perpetually true, necessitating a forced termination of the program's execution.

CHAPTER 3　PYTHON FOR LLMS

For Loops in Python

For loops facilitate iterative processing, allowing for the orderly iteration over structures like lists, strings, or arrays. Python employs a "for in" loop, akin to the foreach loop found in various other programming languages.

Syntax:

```
for iterator_var in sequence:
    statements(s)
```

Example:

```
n = 5
for i in range(0, n):
    print(i)
```

Output:

```
0
1
2
3
4
```

Note　range() function and loops

When using the range() function in conjunction with for loops in Python, it's important to remember that the function generates numbers up to, but not including, the end index. This means if you use range() to iterate up to 5, the loop will execute with the numbers 0 through 4, resulting in a total of 5 iterations. The end index (in this case, 5) is exclusive.

Else Statement with for Loop

Similar to the "while" loop, the "for" loop in Python can be paired with an "else" statement. However, since the "for" loop doesn't terminate based on a conditional expression but rather completes its iteration over a sequence, **the "else" block is executed right after the "for" loop concludes.**

Example:

```
for i in range(3):
    print(f"i is {i}")
else:
    print("Loop completed without break")
```

In this example, the for loop iterates over a sequence generated by range(3), which produces the numbers 0, 1, and 2. For each iteration, the value of i is printed out. Once the loop has iterated over all items in the sequence (i.e., after printing 0, 1, and 2), the loop naturally concludes, and control passes to the else block. The else block then executes, printing "Loop completed without break".

Nested Loops in Python

The Python programming language supports the inclusion of one loop within another, a concept referred to as nested loops. The following examples are provided to demonstrate this concept in action.

Syntax:

For loop:

```
for iterator_var in sequence:
  for iterator_var in sequence:
      statements(s)
  statements(s)
```

While loop:

```
while expression:
   while expression:
       statement(s)
   statement(s)
```

Example:

```python
# Outer loop
for i in range(3):  # Will iterate over 0, 1, 2
    # Inner loop
    for j in range(2):  # Will iterate over 0, 1
        print(f"i = {i}, j = {j}")
```

In this example, there are two loops: an outer loop and an inner loop. The outer loop iterates through a range of numbers from 0 to 2 (inclusive), and for each iteration of the outer loop, the inner loop iterates through a range of numbers from 0 to 1 (inclusive).

The outer loop starts with i = 0. Then, the inner loop begins its execution, iterating with j taking values 0 and then 1. For each iteration of the inner loop, it prints the current values of i and j.

After the inner loop completes its iterations for j = 0 and j = 1, control returns to the outer loop, incrementing i to the next value.

This process repeats until the outer loop has completed all its iterations (for i = 0, i = 1, and i = 2).

The result is a series of prints showing each combination of i and j, demonstrating how nested loops can be used to generate or iterate over a Cartesian product of two ranges. This pattern is commonly used in scenarios requiring iteration over multiple dimensions, such as processing the cells in a 2D matrix or grid.

Output:

```
i = 0, j = 0
i = 0, j = 1
i = 1, j = 0
i = 1, j = 1
i = 2, j = 0
i = 2, j = 1
```

Loop Control Statements

Control statements in loops alter the normal flow of execution. Exiting a scope results in the destruction of all automatically created objects within that scope. Python offers several control statements for this purpose:

- **Continue Statement:** This statement redirects the flow back to the start of the loop, effectively skipping the remainder of the loop's body for the current iteration.

- **Break Statement:** Utilizing this statement exits the loop entirely, transferring control to the statement immediately following the loop.

- **Pass Statement:** The pass statement is used in Python to define a syntactical placeholder, allowing for the creation of empty loops, as well as placeholders within control structures, functions, and classes, without affecting the flow of execution.

Example demonstrating continue, break, and pass:

```
for num in range(1, 10):   # Loop from 1 to 9
    if num % 2 == 0:
        continue   # Skip the rest of the loop for even numbers
    if num == 5:
        pass   # Do nothing for num == 5, placeholder for future code
    if num == 7:
        break   # Exit the loop when num is 7
    print(num)
```

Python Data Structures: Lists, Sets, Tuples, Dictionaries

Python offers a variety of built-in data structures that provide versatile ways to store and manipulate data:

- Lists are ordered collections that allow duplicate elements and support indexing.

- Sets are unordered collections that automatically remove duplicates and provide fast membership testing.

- Tuples are immutable, ordered collections ideal for fixed data sequences.

- Dictionaries are key-value pairs, allowing efficient data retrieval based on unique keys.

Each data structure has distinct characteristics and use cases, making Python a powerful and flexible language for managing data.

What Is a Data Structure?

Data organization, management, and storage play a crucial role in making data access and modification more efficient. **Data structures** provide a framework for arranging data in a manner that facilitates storing data collections, establishing relationships among them, and efficiently executing operations on them.

Python's Built-In Data Structures

Python inherently supports a variety of data structures that allow for effective data storage and retrieval. These include **Lists, Dictionaries, Tuples, and Sets**, each offering unique capabilities for data manipulation.

Custom Data Structures in Python

Python also empowers its users to design their own data structures, giving them complete control over their functionality. Common data structures such as **Stacks, Queues, Trees, and Linked Lists**, familiar to those versed in other programming languages, can be implemented in Python as well. With an understanding of the available data structures, we can now proceed to explore how to implement and utilize these structures in Python.

CHAPTER 3   PYTHON FOR LLMS

Built-In Data Structures

Lists

Lists are versatile data structures in Python that allow you to store a sequence of items of various data types. Each item in a list is assigned a unique index, which is used to access the element. Indexing starts at 0 for the first element, known as positive indexing, and there's also negative indexing, beginning at -1, to access elements starting from the end of the list backward.

Creating Lists

You can create a list by enclosing your elements within square brackets. Leaving the brackets empty will result in an empty list.

Example:

```
my_list = [1, "Hello", 3.14]
Creating an empty list
empty_list = []

print("My list:", my_list)
print("Empty list:", empty_list)
```

Output:

```
My list: [1, 'Hello', 3.14]
Empty list: []
```

Adding Elements

To add elements to a list, Python provides methods such as the following:

- The `append()` method adds its argument as a single element to the end of a list.

- The `extend()` method unfolds its argument, adding each element to the list individually.

- The `insert()` method inserts a given element at a specified index, thereby increasing the list's length.

Example:

```python
# Start with an empty list
fruits = []

# Adding elements to the list using append()
fruits.append("Apple")
fruits.append("Banana")
fruits.append("Cherry")

print("Fruits List:", fruits)
```

Output:

```
Fruits List: ['Apple', 'Banana', 'Cherry']
```

Deleting Elements

Elements can be removed from a list using several techniques:

- The del statement removes elements by index but does not return them.
- The pop() method removes and returns an element at a given index.
- To remove an element by value, you use the remove() method.

Example:

```python
# Creating a list of numbers
numbers = [10, 20, 30, 40, 50]

# Deleting an element by index using del
del numbers[1]  # Deletes 20, which is at index 1
print("After del:", numbers)

# Deleting an element by value using remove()
numbers.remove(30)  # Removes the first occurrence of 30
print("After remove():", numbers)

# Deleting an element and returning it using pop()
popped_element = numbers.pop(2)  # Pops the element at index 2 (which is now 50)
```

```
print("After pop():", numbers)
print("Popped Element:", popped_element)
```

Output:

```
After del: [10, 30, 40, 50]
After remove(): [10, 40, 50]
After pop(): [10, 40]
Popped Element: 50
```

Accessing Elements

Accessing elements in a list is straightforward and similar to accessing characters in a string; you use the index to obtain the desired element.

Example:

```
# Creating a list of fruits
fruits = ["Apple", "Banana", "Cherry", "Date", "Elderberry"]

# Accessing elements by positive indexing
print("First fruit:", fruits[0])  # Apple
print("Third fruit:", fruits[2])  # Cherry

# Accessing elements by negative indexing
print("Last fruit:", fruits[-1])  # Elderberry
print("Second to last fruit:", fruits[-2])  # Date

# Accessing a slice of the list
slice_of_fruits = fruits[1:4]  # From index 1 (inclusive) to index 4 (exclusive)
print("Slice of fruits:", slice_of_fruits)  # ['Banana', 'Cherry', 'Date']
```

In this example:

- We create a list named fruits with five elements.
- We access individual elements using both positive and negative indexes. Positive indexing starts from the beginning of the list (0), and negative indexing starts from the end of the list (-1).

- We also demonstrate how to access a slice (subsection) of the list using the slicing syntax list[start:stop], where start is the index to begin the slice (inclusive) and stop is the index to end the slice (exclusive). This returns a new list containing the specified section.

Additional List Operations

Lists come with a suite of useful methods for manipulation and inquiry:

- **len()** returns the number of items in a list.
- **index()** searches for a given value and returns the index of its first occurrence.
- **count()** counts the occurrences of a given value within the list.
- Both **sorted()** and **sort()** are used for sorting lists. sorted() returns a new sorted list, leaving the original list unchanged, while sort() sorts the list in place.

Example:

```
my_numbers = [5, 2, 8, 10, 25, 10]
print(len(my_numbers))           # Find the length of the list
print(my_numbers.index(10))      # Find the index of the first occurrence of
                                   an element
print(my_numbers.count(10))      # Find the count of a specific element
print(sorted(my_numbers))        # Print a sorted version of the list
                                   without changing the original
my_numbers.sort(reverse=False)   # Sort the original list in ascending order
print(my_numbers)
```

Output:

```
6
3
2
[2, 5, 8, 10, 10, 25]
[2, 5, 8, 10, 10, 25]
```

Understanding these fundamentals of lists will enhance your ability to manage and manipulate data effectively in Python.

Dictionaries in Python

Dictionaries in Python are collections that store data as key-value pairs, akin to a phone directory where each name (key) is associated with a phone number (value). Keys are unique identifiers that map to values, allowing for efficient data retrieval. This structure is akin to looking up a name in a phone book to find the corresponding number.

Creating a Dictionary

You can create dictionaries by enclosing key-value pairs within curly braces {}, or by using the `dict()` constructor. Each key-value pair is added to the dictionary in this manner.

Example:

```python
# Creating a dictionary with key-value pairs
person_info = {"name": "John Doe", "age": 30, "city": "New York"}

# Creating an empty dictionary
empty_dict = {}

print("Person Information:", person_info)
print("Empty Dictionary:", empty_dict)
```

Output:

```
Person Information: {'name': 'John Doe', 'age': 30, 'city': 'New York'}
Empty Dictionary: {}
```

Modifying and Adding Key-Value Pairs

To modify an existing entry, you reference the key and assign a new value to it. Adding a new key-value pair is as simple as assigning a value to a new key in the dictionary.

Example:

```
# Initial dictionary
car = {"make": "Ford", "model": "Mustang", "year": 1964}

# Modifying an existing key-value pair
car["year"] = 2020

# Adding a new key-value pair
car["color"] = "blue"

print("Updated Car Dictionary:", car)
```

Output:

```
Updated Car Dictionary: {'make': 'Ford', 'model': 'Mustang', 'year': 2020, 'color': 'blue'}
```

Removing Key-Value Pairs

- The pop() method removes a key-value pair based on the key and returns the value of the removed pair.
- The popitem() method removes and returns the last key-value pair as a tuple.
- The clear() method empties the entire dictionary, removing all its contents.

Example:

```
# Initial dictionary
book = {"title": "The Great Gatsby", "author": "F. Scott Fitzgerald", "year": 1925}

# Removing a key-value pair using pop()
removed_year = book.pop("year")
print("Removed Year:", removed_year)
print("Book Dictionary after pop():", book)

# Removing a key-value pair using del
del book["author"]
print("Book Dictionary after del:", book)
```

```
# Removing the last inserted key-value pair using popitem()
removed_item = book.popitem()
print("Removed Item:", removed_item)
print("Book Dictionary after popitem():", book)

# Clearing the entire dictionary using clear()
book.clear()
print("Book Dictionary after clear():", book)
```

Output:

```
Removed Year: 1925
Book Dictionary after pop(): {'title': 'The Great Gatsby', 'author': 'F. Scott Fitzgerald'}
Book Dictionary after del: {'title': 'The Great Gatsby'}
Removed Item: ('title', 'The Great Gatsby')
Book Dictionary after popitem(): {}
Book Dictionary after clear(): {}
```

Accessing Elements

Elements are accessed through their keys. You can retrieve a value by referencing its key directly or by using the get() method, which returns the value associated with a given key.

Example:

```
# Creating a dictionary
student_info = {
    "name": "Alice",
    "age": 25,
    "grade": "A"
}

# Accessing elements directly by keys
name = student_info["name"]
age = student_info["age"]

print("Name:", name)
print("Age:", age)
```

```
# Accessing elements using get() method
grade = student_info.get("grade")

print("Grade:", grade)
```

Output:

```
Name: Alice
Age: 25
Grade: A
```

Other Functions

Dictionaries offer several methods for interacting with the data they contain:

- `keys()` method returns a view of the dictionary's keys.
- `values()` method provides a view of the values.
- `items()` method returns a view of the key-value pairs as tuples.

This overview introduces the basic operations and methods available for dictionaries in Python, showcasing their flexibility and power in organizing and accessing data.

Example:

```
# Creating a dictionary
student_info = {
    "name": "Alice",
    "age": 25,
    "grade": "A"
}

# Using keys() method to get keys
keys = student_info.keys()
print("Keys:", keys)

# Using values() method to get values
values = student_info.values()
print("Values:", values)
```

```python
# Using items() method to get key-value pairs
items = student_info.items()
print("Items:", items)

# Iterating through key-value pairs using items()
print("Iterating through key-value pairs:")
for key, value in items:
    print(key, ":", value)
```

Output:

```
Keys: dict_keys(['name', 'age', 'grade'])
Values: dict_values(['Alice', 25, 'A'])
Items: dict_items([('name', 'Alice'), ('age', 25), ('grade', 'A')])
Iterating through key-value pairs:
name : Alice
age : 25
grade : A
```

Tuples in Python

Tuples in Python resemble lists in many ways, except for one crucial difference: once data is added to a tuple, it cannot be altered or modified. There is, however, an exception: when the data contained within a tuple is mutable, it can be changed.

Creating a Tuple

You can create a tuple using parentheses () or by using the tuple() function.

Example:

```python
# Creating a tuple using parentheses
fruits_tuple = ("Apple", "Banana", "Cherry", "Date")

# Creating a tuple using the tuple() constructor
colors_tuple = tuple(("Red", "Green", "Blue"))

print("Fruits Tuple:", fruits_tuple)
print("Colors Tuple:", colors_tuple)
```

Output:

```
Fruits Tuple: ('Apple', 'Banana', 'Cherry', 'Date')
Colors Tuple: ('Red', 'Green', 'Blue')
```

Accessing Elements

Accessing elements in a tuple is identical to accessing values in lists.

Example:

```python
# Creating a tuple
fruits_tuple = ("Apple", "Banana", "Cherry", "Date")

# Accessing elements by index
first_fruit = fruits_tuple[0]
second_fruit = fruits_tuple[1]

print("First Fruit:", first_fruit)
print("Second Fruit:", second_fruit)

# Accessing elements using negative indexing
last_fruit = fruits_tuple[-1]
second_last_fruit = fruits_tuple[-2]

print("Last Fruit:", last_fruit)
print("Second Last Fruit:", second_last_fruit)
```

Output:

```
First Fruit: Apple
Second Fruit: Banana
Last Fruit: Date
Second Last Fruit: Cherry
```

Appending Elements

To append values to a tuple, you can use the + operator, which allows you to concatenate another tuple onto it.

CHAPTER 3 PYTHON FOR LLMS

Example:

```
# Creating two tuples
tuple1 = (1, 2, 3)
tuple2 = (4, 5, 6)

# Appending elements by creating a new tuple
appended_tuple = tuple1 + tuple2

print("Tuple 1:", tuple1)
print("Tuple 2:", tuple2)
print("Appended Tuple:", appended_tuple)
```

Output:

```
Tuple 1: (1, 2, 3)
Tuple 2: (4, 5, 6)
Appended Tuple: (1, 2, 3, 4, 5, 6)
```

Other Functions

Functions available for tuples are similar to those for lists, as tuples share many characteristics with lists in terms of data access and manipulation.

Example:

```
# Creating a tuple with integers and a mutable list
my_tuple = (1, 2, 3, ['hindi', 'python'])

# Modifying an element inside the tuple (a list element)
my_tuple[3][0] = 'english'

# Printing the modified tuple
print("Modified Tuple:", my_tuple)

# Counting the occurrences of an element in the tuple
count_2 = my_tuple.count(2)

# Finding the index of a specific element (the modified list) in the tuple
index_element = my_tuple.index(['english', 'python'])
```

# Printing the count and index results
print("Count of '2' in Tuple:", count_2)
print("Index of ['english', 'python'] in Tuple:", index_element)

Output:

```
Modified Tuple: (1, 2, 3, ['english', 'python'])
Count of '2' in Tuple: 1
Index of ['english', 'python'] in Tuple: 3
```

Sets in Python

Sets in Python are collections of unique and unordered elements. This means that even if data is repeated multiple times, it will be included in the set only once, similar to mathematical sets. Sets support operations akin to those used in arithmetic sets.

Creating a Set

To create a set in Python, you use curly braces "{}". Unlike dictionaries, you only provide values, not key-value pairs.

Example:

```
# Creating a set with unique elements
my_set = {1, 2, 3, 4, 4, 5}

# Printing the set
print("My Set:", my_set)
```

Output:

My Set: {1, 2, 3, 4, 5}

Adding Elements

To add elements to a set, you can use the "add()" function and pass the value you want to add.

Example:

```
# Creating an empty set
my_set = set()
```

```python
# Adding elements to the set using the add() method
my_set.add(1)
my_set.add(2)
my_set.add(3)

# Printing the updated set
print("Updated Set:", my_set)
```

Output:

Updated Set: {1, 2, 3}

Operations on Sets

Various set operations like union and intersection are demonstrated here. These operations allow you to manipulate and combine sets as needed.

Example:

```
# Creating two sets
my_set = {1, 2, 3, 4}
my_set_2 = {3, 4, 5, 6}

# Performing set operations and comparing them using union and | operator
union_result = my_set.union(my_set_2)
union_operator_result = my_set | my_set_2

# Performing set operations and comparing them using intersection and & operator
intersection_result = my_set.intersection(my_set_2)
intersection_operator_result = my_set & my_set_2

# Performing set operations and comparing them using difference and - operator
difference_result = my_set.difference(my_set_2)
difference_operator_result = my_set - my_set_2

# Performing set operations and comparing them using symmetric_difference and ^ operator
symmetric_difference_result = my_set.symmetric_difference(my_set_2)
symmetric_difference_operator_result = my_set ^ my_set_2
```

CHAPTER 3   PYTHON FOR LLMS

```
# Clearing the first set
my_set.clear()

# Printing the results with added comments
print("Union Result:", union_result, 'is equivalent to', union_
operator_result)
print("Intersection Result:", intersection_result, 'is equivalent to',
intersection_operator_result)
print("Difference Result:", difference_result, 'is equivalent to',
difference_operator_result)
print("Symmetric Difference Result:", symmetric_difference_result, 'is
equivalent to', symmetric_difference_operator_result)
print("Cleared Set:", my_set)   # The set is now empty
```

Output:

```
Union Result: {1, 2, 3, 4, 5, 6} is equivalent to {1, 2, 3, 4, 5, 6}
Intersection Result: {3, 4} is equivalent to {3, 4}
Difference Result: {1, 2} is equivalent to {1, 2}
Symmetric Difference Result: {1, 2, 5, 6} is equivalent to {1, 2, 5, 6}
Cleared Set: set()
```

In this example:

- **Union:** The union() function combines the data from both sets, creating a new set that contains all unique elements from both sets.

- **Intersection:** The intersection() function finds the data that is common to both sets. It returns a new set containing only the elements that exist in both sets.

- **Difference:** The difference() function removes the data that is common to both sets from the first set and outputs a new set containing only the data present in the first set but not in the second set.

- **Symmetric Difference:** The symmetric_difference() function is similar to the difference() function but with a twist. It removes the data that is common to both sets from both sets and outputs a new set containing the data that remains unique in each set.

CHAPTER 3   PYTHON FOR LLMS

These functions allow you to perform various set operations and manipulate sets to get specific subsets of data as needed.

Regular Functions and Lambda Functions
What Is a Function in Python?

In programming, functions are utilized to group together a series of steps that you either need to execute multiple times or that are sufficiently complex to warrant encapsulation in a distinct subprogram for invocation as required. Essentially, a function is a segment of code designed to perform a particular action. Depending on the nature of this action, a function may require several inputs to operate, and upon completion, it has the capability to return outcomes, either singular or multiple.

Python distinguishes three categories of functions:

- **Built-in functions**, like help() for requesting assistance, min() for determining the smallest value, and print() for outputting an object to the terminal, among others. A more comprehensive list of these functions is available.

- **User-Defined Functions (UDFs)**, which are custom functions crafted by users to facilitate their tasks.

- **Anonymous functions**, often referred to as lambda functions, are unique in that they are defined without employing the conventional def keyword.

Creating a User-Defined Function (UDF) in Python involves four primary steps:

1. Start by using the "def" keyword to signal the creation of a function, followed by the chosen name for the function.

2. Specify any parameters the function requires, placing them inside the parentheses next to the function name. Conclude this line with a colon.

3. Incorporate the instructions that you want the function to carry out.

CHAPTER 3 PYTHON FOR LLMS

4. Conclude the function with a "return" statement if you intend for the function to produce a result. Omitting the return statement means the function will yield a "None" object by default.

Example:

```
def add_numbers(a, b):
    return a + b

# Example of calling the function
result = add_numbers(3, 5)
print(result)  # This will print 8
```

Output:

8

The return Statement

The return statement in Python is used within a function to exit it and pass back a value to the caller. A function can return a value, including data types like integers, floats, strings, lists, tuples, dictionaries, or even other functions and objects. If a function doesn't explicitly end with a return statement, Python automatically returns None, indicating that the function doesn't provide any specific value.

The `return` statement can be used to

- Return a specific result calculated or processed within the function
- Exit the function at a certain point before the end of its block of code
- Pass control back to the point in the program where the function was called, optionally passing data back to that point

Return or Print in a Function

In Python, "`return`" and "`print`" within a function serve different purposes:

- `return` sends a value back to the caller and **exits the function**. It allows the function to output a result, which can then be used elsewhere in your program. The returned value can be stored in a variable, used in expressions, or passed to other functions.
- `print()`, on the other hand, simply displays a value to the console. It does not exit the function or send any value back to the caller. "print" is used for logging or debugging purposes, allowing developers to see outputs without affecting the flow or output of the function.

In short, use "`return`" when you want to output a value from a function and use it further, and use "`print`" when you want to display something to the console without affecting the function's output to its caller.

Methods vs. Functions

A method is a type of function that is associated with a class and is accessed via an instance or object of that class. On the other hand, a function is a stand-alone entity without the requirement of being linked to a class. Therefore, while every method qualifies as a function, the reverse does not hold true for all functions.

Take this scenario as an illustration: initially, you create a function named plus(), followed by the establishment of a Summation class that encompasses a sum() method. To utilize the sum() method that is integrated into the Summation class, it's imperative to instantiate an object or instance of this class. Let's proceed to construct such an object.

How to Call a Function in Python

To call a function in Python, you simply use the function's name followed by parentheses. If the function requires arguments, you place them inside the parentheses, separated by commas. Here's the basic syntax:

Syntax:

```
function_name(arguments)
```

For example:

```
result = add_numbers(5, 3)
```

This executes the "add_numbers" function with "5" and "3" as its arguments and stores the return value in the variable "result". If the function does not require any arguments, you still need to use parentheses, but leave them empty:

```
function_name()
```

This syntax is straightforward and is the same for both built-in functions and user-defined functions.

Function Arguments in Python

In Python, function arguments are values passed to a function when it is called. These arguments are used by the function to perform operations or produce results based on the input. Python supports various types of arguments, making functions highly flexible in handling inputs. Here are the main types.

Positional Arguments

These are the most common type of arguments, where the order in which the arguments are passed matters. The number of positional arguments in the call must match the number expected by the function definition.

Example:

```
def add(a, b):
    return a + b

# Calling function with positional arguments
result = add(2, 3)   # 2 and 3 are positional arguments
print(result)
```

Output:

5

Keyword Arguments

These are arguments passed to a function by explicitly specifying the name of the parameter and the value. Keyword arguments can be listed in any order when calling the function. This makes your code more readable and allows you to call functions with the parameters in a different order than they were defined.

Example:

```
def greet(name, message):
    return f"{message}, {name}!"

# Calling function with keyword arguments
greeting = greet(message="Hello", name="Alice")
```

Default Arguments

A function can have default values for parameters. If the caller does not provide a value for such an argument, the function uses the default value. Default arguments make certain parameters optional.

Example:

```
def log(message, level='INFO'):
    print(f"[{level}] {message}")

# Calling function without specifying level
log("User logged in")  # Uses default level INFO
```

Variable-Length Arguments (*args and **kwargs)

- *args allows a function to accept an arbitrary number of positional arguments. It is used when you are not sure how many arguments might be passed to your function.

- **kwargs allows for an arbitrary number of keyword arguments. It is used when you want to handle named arguments that you have not explicitly defined in advance.

Example:

```
def make_sentence(*words, **punctuation):
    sentence = ' '.join(words) + punctuation.get('mark', '.')
    return sentence.capitalize()

# Using *args for words and **kwargs for punctuation
sentence = make_sentence('hello', 'world', mark='!')
```

Function arguments increase the flexibility and reusability of your functions, allowing you to write more generalized and powerful code.

Anonymous Functions in Python

Anonymous functions in Python, also known as lambda functions, are functions defined without a name. They are created using the "lambda" keyword, which is why they are often referred to as lambda functions. Anonymous functions are typically used for short, simple operations that are easier to read when written inline, particularly as arguments to higher-order functions (functions that take other functions as inputs).

The basic syntax of an anonymous function is

```
lambda arguments: expression
```

This syntax allows the lambda function to take any number of arguments, but it can only have one expression. The expression is evaluated and returned when the lambda function is called.

Characteristics of Lambda Functions

- **Single Expression**: Unlike standard functions defined with "def", which can consist of multiple expressions and statements, a lambda function is limited to a single expression.

- **Automatically Returned**: The result of the expression is automatically returned by the lambda function.

- **No Need for a Return Statement**: Unlike normal functions, there's no need to explicitly return the result; the expression's outcome is automatically the return value.

- **Versatile Usage**: Lambda functions can be used anywhere a function is expected, and they are often used in combination with functions like "map()", "filter()", and "reduce()", as well as in GUI event handlers and other short callback functions.

Example:

```
square = lambda x: x * x
print(square(5))   # Output: 25
```

Output:

25

And here's an example of using a lambda function with the "filter()" function to filter out even numbers from a list:

```
numbers = [1, 2, 3, 4, 5, 6]
even_numbers = list(filter(lambda x: x % 2 == 0, numbers))
print(even_numbers)   # Output: [2, 4, 6]
```

Output:

[2,4,6]

Lambda functions are particularly useful for simple operations that can be expressed in a single line. However, for complex operations, it's recommended to use named functions for the sake of clarity and readability.

Summary

Python is a general-purpose, high-level programming language, an ideal choice for scripting and rapid application development across various domains and platforms. In this chapter, we reviewed Python's syntax and semantics and noted its similarity to other programming languages while highlighting its unique features. In the next chapter, we'll see how Python supports various programming approaches, including structured, object-oriented, and functional programming.

CHAPTER 4

Python and Other Programming Approaches

Python's rich NLP ecosystem, active community support, readability, integration with data science and machine learning tools, availability of pre-trained models, open source nature, wide adoption, versatility, and cross-platform compatibility make it an ideal choice for natural language processing tasks. Its popularity and robust ecosystem have made Python the de facto language for NLP research and development.

Object-Oriented Programming in Python

Object-Oriented Programming (OOP) in Python is a coding approach that models complex issues using objects. This approach is grounded in a theoretical framework that guides problem-solving strategies.

In the context of OOP, this involves employing a suite of principles and design patterns to address challenges through the lens of objects. In Python, an object is an encapsulation of data (known as attributes) and functions (referred to as methods) into a single entity. Objects can be likened to tangible entities in the real world.

Attributes, in this scheme, are typically nouns, signifying the object's properties, while methods are actions, expressed as verbs, that define what the object can do.

This division into data storage and operational functionality forms the core of Object-Oriented Programming. It involves crafting objects that not only hold information but also embody specific behaviors.

CHAPTER 4   PYTHON AND OTHER PROGRAMMING APPROACHES

Why Do We Use Object-Oriented Programming in Python?

We use Object-Oriented Programming (OOP) in Python to organize code into reusable, modular units called objects, which combine data and functionality. This approach simplifies complex software development, making code more understandable, maintainable, and scalable.

Object-Oriented Programming

- Enhances code reuse through **inheritance**.
- Allows for flexible code interaction via **polymorphism**.
- Improves problem-solving by breaking down problems into manageable objects.
- Promotes **data encapsulation**, ensuring secure and structured code management. This methodology aligns well with Python's philosophy of clear, logical code for both small and large-scale projects.

OOP vs. Structured Programming

- **Maintenance:** OOP is generally more maintainable compared to structured programming, which can be more challenging to maintain.
- **Code Duplication:** OOP follows the Don't Repeat Yourself (DRY) principle, minimizing code repetition, whereas structured programming may involve code being duplicated across multiple locations.
- **Code Reusability:** In OOP, small, reusable code segments are common, unlike structured programming, where larger blocks of code are centralized.
- **Programming Model:** OOP utilizes an object-based model, while structured programming relies on a sequential block of code model.
- **Debugging:** Debugging is typically more straightforward in OOP due to its modular nature, whereas it can be more complex in structured programming.

- **Learning Curve:** OOP comes with a steeper learning curve, whereas structured programming is often seen as easier for beginners to grasp.

- **Project Suitability:** OOP is preferred for complex, large-scale projects, while structured programming is more suited to simpler, smaller programs.

Everything Is an Object in Python

Here's a little insider knowledge: you've been immersed in Object-Oriented Programming (OOP) during your entire journey with Python, perhaps without even realizing it. In Python, regardless of the programming paradigm you're working within, you're engaging with objects for virtually every operation.

This stems from Python's core principle: **everything in Python is treated as an object**. Recall what constitutes an object: In Python, an object encapsulates both data (known as attributes) and functionalities (referred to as methods). This concept perfectly aligns with every data type in Python.

Take a string, for example; it's essentially a collection of data (the characters within it) coupled with a set of functionalities or behaviors (methods like upper(), lower(), and so on). This principle equally applies to other data types like integers, floats, booleans, lists, and dictionaries.

Example:

```
text = "Hello, Python!"
print(text.upper())   # Using the upper() method
```

Output:

```
HELLO PYTHON
```

Attributes and Methods

Before we proceed, it's beneficial to revisit what we mean by attributes and methods:

- **Attributes** are akin to internal variables housed within objects, holding data relevant to the object.

- **Methods** are similar to functions, except they are bound to the objects themselves, offering specific behaviors or actions that can be performed with or on the object's data.

Your First Python Object

Let's consider a simple example of an object in Python by creating a "Book" class and then instantiating it to create a "Book" object. This example will illustrate how Python treats everything as an object, encompassing both attributes (data) and methods (functionality).

Defining the 'Book' Class

First, we define a class named "Book", which is like a blueprint for creating "Book" objects. This class will have attributes to store data such as the title and author of the book and a method to display this information.

```
class Book:
    def __init__(self, title, author):
        self.title = title
        self.author = author

    def display_info(self):
        print(f"Book Title: {self.title}\nAuthor: {self.author}")
```

Explanation:

- **__init__** is a special method in Python, known as the constructor. It initializes new objects from the class, setting up their initial state. In this case, it takes "title" and "author" as parameters and assigns them to the object's attributes ("self.title" and "self.author"). It's called Python each time we instantiate an object.

- **display_info** is a method defined to perform an operation related to the object, which is displaying the book's title and author.

Creating and Using a "Book" Object

Now, let's create an actual "Book" object using the "Book" class blueprint and then use its method.

```
# Creating a Book object
my_book = Book("The Catcher in the Rye", "J.D. Salinger")

# Using the display_info method of the Book object
my_book.display_info()
```

This will output

```
Book Title: The Catcher in the Rye
Author: J.D. Salinger
```

The "my_book" instance is an object of the "Book" class. It encapsulates data (the title and author attributes) and provides functionality through a method ("display_info"). This object is a practical embodiment of OOP principles in Python, demonstrating how data and related behaviors are bundled together. Each "Book" object can hold different data, showcasing the power of objects in managing state and behavior in a structured way.

Object-Oriented Programming (OOP) in Python Is Founded on Four Fundamental Concepts

Abstraction

Abstraction involves concealing the complex realities of a system to simplify its interaction for the user, who could be the end user or other developers. It's akin to knowing how to operate a smartphone without understanding the intricate details of its internal processes, or using Python to develop applications without needing to grasp its underlying mechanics. In programming, abstraction allows for the distillation of common functionalities into classes, making the management of objects and their interactions more straightforward.

Example:

```python
from abc import ABC, abstractmethod

# Define an abstract base class (ABC)
class Shape(ABC):

    @abstractmethod
    def area(self):
        pass

# Create a concrete class that inherits from Shape
class Circle(Shape):
```

```python
    def __init__(self, radius):
        self.radius = radius

    def area(self):
        return 3.1415 * self.radius ** 2

# Create a concrete class that inherits from Shape
class Rectangle(Shape):

    def __init__(self, width, height):
        self.width = width
        self.height = height

    def area(self):
        return self.width * self.height

# Instantiate objects and calculate areas
circle = Circle(5)
rectangle = Rectangle(4, 6)

print("Circle Area:", circle.area())  # Output: Circle Area: 78.53750000000001
print("Rectangle Area:", rectangle.area())  # Output: Rectangle Area: 24
```

Output:

Circle Area: 78.53750000000001
Rectangle Area: 24

Inheritance

Inheritance is the mechanism by which classes can derive characteristics and behaviors from pre-existing classes. This principle supports the Don't Repeat Yourself (DRY) philosophy, enabling the reuse of code by incorporating shared attributes and methods into base classes (superclasses). This concept mirrors biological inheritance, where subclasses inherit traits and behaviors from their parents (superclasses), showcasing shared attributes and potentially shared methods.

Example:

```python
# Parent class (superclass)
class Animal:
    def __init__(self, name):
        self.name = name

    def speak(self):
        pass  # Placeholder for the speak method

# Child class (subclass) inheriting from Animal
class Dog(Animal):
    def speak(self):
        return f"{self.name} says Woof!"

# Child class (subclass) inheriting from Animal
class Cat(Animal):
    def speak(self):
        return f"{self.name} says Meow!"

# Create instances of the child classes
dog = Dog("Buddy")
cat = Cat("Whiskers")

# Call the speak method for each instance
print(dog.speak())   # Output: Buddy says Woof!
print(cat.speak())   # Output: Whiskers says Meow!
```

Output:

Buddy says Woof!
Whiskers says Meow!

Polymorphism

Polymorphism permits the tailoring of methods and attributes in subclasses that were initially defined in a superclass, embodying the concept of "many forms." It allows for the creation of methods with the same name but differing implementations across classes. Reflecting on our real-life analogy, children (subclasses) may share a common behavior like getting hungry (a method) from their parents, but the specifics, such as frequency of hunger, can vary, demonstrating polymorphism.

Example:

```python
# Parent class (superclass)
class Animal:
    def __init__(self, name):
        self.name = name

    def speak(self):
        pass  # Placeholder for the speak method

# Child class (subclass) inheriting from Animal
class Dog(Animal):
    def speak(self):
        return f"{self.name} says Woof!"

# Child class (subclass) inheriting from Animal
class Cat(Animal):
    def speak(self):
        return f"{self.name} says Meow!"

# Function that takes an Animal object and calls its speak method
def animal_sound(animal):
    return animal.speak()

# Create instances of the child classes
dog = Dog("Buddy")
cat = Cat("Whiskers")

# Call the animal_sound function with different objects
print(animal_sound(dog))  # Output: Buddy says Woof!
print(animal_sound(cat))  # Output: Whiskers says Meow!
```

Output:

```
Buddy says Woof!
Whiskers says Meow!
```

Encapsulation

Encapsulation ensures the safekeeping of a class's internal data, safeguarding the integrity and privacy of the data within. Although Python does not explicitly support private attributes through syntax, it achieves encapsulation through name mangling and by using **getter and setter methods** for controlled access and modification of data.

Example:

```
class Human:
    def __init__(self, name, age):
        self.__name = name    # Private attribute
        self.__age = age      # Private attribute

    # Getter method for name
    def get_name(self):
        return self.__name

    # Setter method for age
    def set_age(self, age):
        if age > 0 and age < 150:    # Adding validation
            self.__age = age
        else:
            print("Invalid age")

    # Method to display information
    def display_info(self):
        print(f"Name: {self.__name}, Age: {self.__age}")

# Create an instance of the Human class
person = Human("Alice", 30)

# Accessing attributes directly (not recommended)
# This will raise an error: AttributeError: 'Human' object has no attribute '__name'
# print(person.__name)

# Accessing attributes using getter method
print("Name:", person.get_name())    # Output: Name: Alice
```

```
# Setting age using setter method (with validation)
person.set_age(35)
person.display_info()   # Output: Name: Alice, Age: 35
# Setting an invalid age
person.set_age(200)   # Output: Invalid age
```

Output:

```
Name: Alice
Name: Alice, Age: 35
Invalid age
```

Modules and File Handling

Python Modules

A Python module is essentially a file that houses a collection of built-in functions, classes, and variables, each serving a specific purpose. Python boasts numerous modules, each tailored to handle distinct tasks. In this article, we will delve into the realm of Python modules, exploring topics such as creating custom modules, importing modules in Python, utilizing aliases for module names, and more.

Understanding Python Modules

A Python module represents a container for Python definitions and statements, encompassing functions, classes, variables, and even executable code. Organizing related code within a module enhances code comprehension, logical organization, and reusability.

Creating a Python Module

To craft a Python module, you simply write the desired code and save it in a file bearing the .py extension with the name math_operations.

Example:

```
# math_operations.py

# Function to add two numbers
def add(x, y):
    return x + y
```

```
# Function to subtract two numbers
def subtract(x, y):
    return x - y
```

Importing Modules in Python

Python enables us to import functions and classes defined within a module into another module using the import statement within the source file. When the interpreter encounters an import statement, it imports the module if it resides in the search path.

> **Note** Search path
>
> *The search path is essentially a list of directories that the interpreter scans when searching for a module.*

If you keep the previous file we created and create a new one to call it, you should have the following structure:

```
├── calc.py
└── math_operations.py
```

Now, to import the module named "math_operations.py," include the following command at the beginning of your script:

Example:

```
# Import the math_operations module
import math_operations

# Use the functions from the module
result_add = math_operations.add(5, 3)
result_subtract = math_operations.subtract(10, 4)
print("Addition result:", result_add)
print("Subtraction result:", result_subtract)
```

Output:

```
Addition result: 8
Subtraction result: 6
```

> **Note** The dot operator
>
> *This imports the module itself, not its functions or classes. To access the contents of the module, you use the dot (.) operator.*

Python Import Using "from" Statement

Python "from" statement allows for the selective import of specific attributes from a module without importing the entire module.

Example:

```python
# Importing the math module and specifically importing the sqrt function
from math import sqrt

# Using the imported function to calculate the square root
number = 25
result = sqrt(number)

# Displaying the result
print(f"The square root of {number} is {result}")
```

Output:

```
The square root of 25 is 5.0
```

Importing Specific Attributes from a Python Module

Example:

```python
# Importing specific attributes from the built-in "math" library
from math import pi, sqrt

# Using the imported attributes
print(f"The value of pi is approximately: {pi}")

number = 25
result = sqrt(number)
print(f"The square root of {number} is: {result}")
```

Output:

```
The value of pi is approximately: 3.141592653589793
The square root of 25 is: 5.0
```

Importing All Names

The asterisk (*) symbol, when used in conjunction with the import statement, imports all names from a module into the current namespace.

Syntax:

```
from module_name import *
```

The use of "*" has its advantages and disadvantages. It is not recommended to use it if you know precisely what you need from the module. Use it judiciously.

Locating Python Modules

When a module is imported in Python, the interpreter searches for it in various locations. Initially, it checks for built-in modules; if not found, it explores a list of directories defined in the "sys.path" variable. The "sys.path" variable consists of a list of directories where Python searches for the required module.

Example:

```python
import sys
import importlib

# Add a custom directory to the module search path
custom_module_path = "/path/to/custom_module_directory"
sys.path.append(custom_module_path)

# Attempt to import a module from the custom directory
module_name = "custom_module"

try:
    custom_module = importlib.import_module(module_name)
    print(f"Successfully imported {module_name} from {custom_module.__file__}")
except ImportError:
    print(f"Failed to import {module_name}")
```

```
# Print the current module search path
print("Current module search path:")
for path in sys.path:
    print(path)
```

Output on my computer:

```
Failed to import custom_module
Current module search path:
C:\Users\didog\OneDrive\Desktop\test\test
C:\Users\didog\OneDrive\Desktop\test\test
C:\Program Files\JetBrains\PyCharm 2023.2\plugins\python\helpers\pycharm_display
C:\Users\didog\AppData\Local\Programs\Python\Python311\python311.zip
C:\Users\didog\AppData\Local\Programs\Python\Python311\DLLs
C:\Users\didog\AppData\Local\Programs\Python\Python311\Lib
C:\Users\didog\AppData\Local\Programs\Python\Python311
C:\Users\didog\OneDrive\Desktop\test\test\.venv
C:\Users\didog\OneDrive\Desktop\test\test\.venv\Lib\site-packages
C:\Program Files\JetBrains\PyCharm 2023.2\plugins\python\helpers\pycharm_matplotlib_backend
/path/to/custom_module_directory
```

Renaming Python Modules

It is possible to rename a module while importing it by using the "as" keyword.

Syntax:

```
import Module_name as Alias_name
```

Python Built-In Modules

Python offers a wide range of built-in modules that provide various functionalities and features. Some of the most popular built-in modules in Python include the following:

- **math:** Provides mathematical functions and constants like `"sqrt"`, `"sin"`, `"cos"`, `"pi"`, and more.
- **os:** Allows interaction with the operating system, enabling tasks like file and directory operations, environment variables, and process management.

- **sys:** Offers access to Python interpreter variables and functions, such as command-line arguments and system-specific settings.

- **datetime:** Facilitates working with dates and times, allowing you to create, manipulate, and format dates and time intervals.

- **random:** Provides functions for generating random numbers and performing random sampling.

- **json:** Enables encoding and decoding JSON (JavaScript Object Notation) data, making it easy to work with JSON files and web APIs.

- **re:** Supports regular expressions for pattern matching and text manipulation.

- **collections:** Offers specialized data structures like "Counter", "defaultdict", and "namedtuple" for more advanced data handling.

- **urllib:** Provides tools for working with URLs and fetching data from web resources, often used in web scraping and web requests.

- **csv:** Offers utilities for reading and writing CSV (Comma-Separated Values) files, commonly used for data storage and exchange.

- **sqlite3:** Allows interaction with SQLite databases, making it suitable for lightweight database operations.

- **argparse:** Simplifies the creation of command-line interfaces (CLIs) by parsing command-line arguments and generating help messages.

- **os.path:** Provides methods for manipulating file paths and checking file and directory existence.

- **logging:** Provides a flexible and configurable logging framework for debugging and monitoring your code.

- **socket:** Offers network communication capabilities, including creating and interacting with sockets for network connections.

- **multiprocessing:** Facilitates parallel and concurrent programming by providing a higher-level interface to process management.

These are just a few of the popular built-in modules in Python. Python's extensive standard library includes many more modules, each serving specific purposes and making Python a versatile language for a wide range of applications.

Python File Handling

Python offers robust file handling capabilities that enable users to interact with files, encompassing tasks such as **reading, writing, and a range of other file-related operations**. While the concept of file handling exists in various programming languages, Python distinguishes itself by providing a straightforward and concise implementation. When it comes to file manipulation, Python treats files differently based on whether they are in text or binary format, a crucial distinction to keep in mind.

In Python, each line of code consists of a sequence of characters, effectively forming a text file. It's worth noting that each line in a text file is terminated by a special character known as the End of Line (EOL) character, which can be a comma {,} or a newline character. This EOL character marks the conclusion of the current line and signals to the interpreter that a new line is about to begin. Let's delve into the essentials of reading and writing files in Python.

Python File Opening

Before performing any file operations, such as reading or writing, the initial step is to open the file. To achieve this, Python provides the built-in function "open()". However, during the file's opening, it is crucial to specify the mode, which indicates the purpose of opening the file.

Example:
Create and put the file example.txt in the same folder of your .py file.
My example.txt contains the following:

Lorem Ipsum is simply dummy text of the printing and typesetting industry. Lorem Ipsum has been the industry's standard dummy text ever since the 1500s, when an unknown printer took a galley of type and scrambled it to make a type specimen book. It has survived not only five centuries, but also the leap into electronic typesetting, remaining essentially unchanged. It was popularised in the 1960s with the release of Letraset sheets containing Lorem Ipsum passages, and more recently with desktop publishing software like Aldus PageMaker including versions of Lorem Ipsum.

If we follow the tree structure, then the files should be as follows:

```
├── example.txt
└── test.py
```

```python
# Specify the file path (replace with the actual file path)
file_path = "example.txt"

try:
    # Open the file and read its contents
    file = open(file_path, 'r')
    file_contents = file.read()

    # Print the file contents
    print(file_contents)

    # Close the file
    file.close()
except FileNotFoundError:
    print(f"The file '{file_path}' was not found.")
except IOError:
    print(f"An error occurred while reading the file '{file_path}'.")
```

Output:

Lorem Ipsum is simply dummy text of the printing and typesetting industry. Lorem Ipsum has been the industry's standard dummy text ever since the 1500s, when an unknown printer took a galley of type and scrambled it to make a type specimen book. It has survived not only five centuries, but also the leap into electronic typesetting, remaining essentially unchanged. It was popularized in the 1960s with the release of Letraset sheets containing Lorem Ipsum passages, and more recently with desktop publishing software like Aldus PageMaker including versions of Lorem Ipsum.

The following modes are supported:

- r: Opens an existing file for reading.
- w: Opens an existing file for writing. If the file already contains data, it will be overwritten. If the file doesn't exist, it will be created.

- **a**: Opens an existing file for append operation. It won't overwrite existing data.
- **r+**: Allows reading and writing data into the file. Existing data in the file will be overwritten.
- **w+**: Enables writing and reading data. It will overwrite existing data.
- **a+**: Permits appending and reading data from the file. It won't overwrite existing data.

Working in Read Mode

There are multiple approaches to reading from a file in Python. Let's explore how we can retrieve the contents of a file when it's opened in read mode.

In the following example, we'll open the file test.txt with the following content:

Lorem Ipsum is simply dummy text of the printing and typesetting industry. Lorem Ipsum has been the industry's standard dummy text ever since the 1500s, when an unknown printer took a galley of type and scrambled it to make a type specimen book. It has survived not only five centuries, but also the leap into electronic typesetting, remaining essentially unchanged. It was popularised in the 1960s with the release of Letraset sheets containing Lorem Ipsum passages, and more recently with desktop publishing software like Aldus PageMaker including versions of Lorem Ipsum.

Example:

```
# Open a file named "test.txt" in reading mode ('r')
file = open('test.txt', 'r')

# Print each line from the file one by one
for line in file:
    print(line, end='')  # Print each line without adding extra newlines

# Close the file
file.close()
```

Output:

Lorem Ipsum is simply dummy text of the printing and typesetting industry. Lorem Ipsum has been the industry's standard dummy text ever since the 1500s, when an unknown printer took a galley of type and scrambled it to make a type specimen book. It has survived not only five centuries, but also the leap into electronic typesetting, remaining essentially unchanged. It was popularised in the 1960s with the release of Letraset sheets containing Lorem Ipsum passages, and more recently with desktop publishing software like Aldus PageMaker including versions of Lorem Ipsum.

Example:

```
# Python code to demonstrate the read() mode
file = open("test.txt", "r")
print (file.read())
```

Output:

Lorem Ipsum is simply dummy text of the printing and typesetting industry. Lorem Ipsum has been the industry's standard dummy text ever since the 1500s, when an unknown printer took a galley of type and scrambled it to make a type specimen book. It has survived not only five centuries, but also the leap into electronic typesetting, remaining essentially unchanged. It was popularised in the 1960s with the release of Letraset sheets containing Lorem Ipsum passages, and more recently with desktop publishing software like Aldus PageMaker including versions of Lorem Ipsum.

Example:

```
# Python code to illustrate with()
with open("geeks.txt") as file:
    data = file.read()
print(data)
```

Output:

Lorem Ipsum is simply dummy text of the printing and typesetting industry. Lorem Ipsum has been the industry's standard dummy text ever since the 1500s, when an unknown printer took a galley of type and scrambled it to make a type specimen book. It has survived not only five centuries, but also the leap into electronic typesetting, remaining essentially unchanged. It was popularized in the 1960s with the release of Letraset sheets containing Lorem Ipsum passages, and more recently with desktop publishing software like Aldus PageMaker including versions of Lorem Ipsum.

In Python, it's possible to split lines while reading files. The split() function is used to divide a string into parts when a space is encountered by default. However, you can customize the splitting character to be any character of your choice.

Example:

```
# Python code to illustrate split() function
with open("test.txt", "r") as file:
    data = file.readlines()
    for line in data:
        word = line.split()
        print (word)
```

Output:

['Lorem', 'Ipsum', 'is', 'simply', 'dummy', 'text', 'of', 'the', 'printing', 'and', 'typesetting', 'industry.', 'Lorem', 'Ipsum', 'has', 'been', 'the', "industry's", 'standard', 'dummy', 'text', 'ever', 'since', 'the', '1500s,', 'when', 'an', 'unknown', 'printer', 'took', 'a', 'galley', 'of', 'type', 'and', 'scrambled', 'it', 'to', 'make', 'a', 'type', 'specimen', 'book.', 'It', 'has', 'survived', 'not', 'only', 'five', 'centuries,', 'but', 'also', 'the', 'leap', 'into', 'electronic', 'typesetting,', 'remaining', 'essentially', 'unchanged.', 'It', 'was', 'popularised', 'in', 'the', '1960s', 'with', 'the', 'release', 'of', 'Letraset', 'sheets', 'containing', 'Lorem', 'Ipsum', 'passages,', 'and', 'more', 'recently', 'with', 'desktop', 'publishing', 'software', 'like', 'Aldus', 'PageMaker', 'including', 'versions', 'of', 'Lorem', 'Ipsum.']

Creating a File Using the "`write()`" Function

Similar to reading a file in Python, there are various methods to write to a file. Here, we will demonstrate how to write content to a file using the "write()" function in Python.

Example:

```
# Specify the file path (replace with the desired file path)
file_path = "new_file.txt"

try:
    # Open the file in write mode ('w')
    with open(file_path, 'w') as file:
        # Write content to the file
        file.write("This is the first line.\n")
        file.write("This is the second line.\n")
        file.write("This is the third line.\n")

    print(f"File '{file_path}' created and content written successfully.")
except IOError:
    print(f"An error occurred while creating or writing to the file '{file_path}'.")
```

Output:

```
File 'example.txt' created and content written successfully.
```

Now if you open the file, you'll see the following:

```
This is the first line.
This is the second line.
This is the third line.
```

Working in Append Mode

```
# Python code to illustrate append() mode
file = open('test.txt', 'a')
file.write("This will add this line")
file.close()
```

And now in the file, you'll see the following:

```
This is the first line.
This is the second line.
This is the third line.
This will add this line
```

The Powerful Features of Python 3.11

Python 3.11 made its debut on October 24, 2022, boasting increased speed and enhanced user-friendliness. Following 17 months of intensive development, it is now poised for widespread adoption.

Like its predecessors, Python 3.11 introduces a plethora of enhancements and modifications. While a comprehensive list of these can be found in the documentation, this chapter will delve into the most exciting and influential new features.

TypedDicts

TypedDict was specified in PEP 589 and introduced in Python 3.8. On older versions of Python, you can install it from typing_extensions (pip install typing_extensions). In Python 3.11, it is directly imported from typing.

TypedDict or Just a Dict?

- The dict [key: value] type lets you declare uniform dictionary types, where every value has non-defined type, and arbitrary keys are supported.
- But The TypedDict allows us to describe a structured dictionary where the type of each dictionary value depends on the key.

Example:

```
from typing import TypedDict, List

class SalesSummary(TypedDict):
    sales: int
    country: str
    product_codes: List[str]
```

```python
def get_sales_summary() -> SalesSummary:
    return {
        "sales": 1_000,
        "country": "UK",
        "product_codes": ["SUYDT"],
    }
```

TypedDict cannot inherit from both a TypedDict type and a non-TypedDict base class.

Required[ ] and NotRequired[ ]

Example:

```
from typing import TypedDict, NotRequired

class User(TypedDict):
    name: str
    age: int
    married: NotRequired[bool]

marie: User = {'name': 'Marie', 'age': 29, 'married': True}
fredrick : User = {'name': 'Fredrick', 'age': 17}  # 'married' is not required
```

Example:

```
from typing import TypedDict, Required

# 'total=False' means all fields are not required by default
class User(TypedDict, total=False):
    name: Required[str]
    age: Required[int]
    married: bool  # now this is optional

marie: User = {'name': 'Marie', 'age': 29, 'married': True}
fredrick : User = {'name': 'Fredrick', 'age': 17}  # 'married' is not required
thomas: User = {'age': 29, 'married': True} # Will be highlighted because a key is missing!
```

Unary operators + / - / ~ as Required/NotRequired example:

```
class MyThing(TypedDict, total=False):
    req1: +int      # + means a required key, or Required[]
    opt1: str
    req2: +float

class MyThing(TypedDict):
    req1: int
    opt1: -str      # - means a potentially-missing key, or NotRequired[]
    req2: float

class MyThing(TypedDict):
    req1: int
    opt1: ~str      # ~ means a opposite-of-normal-totality key
    req2: float
```

Self Type

Previously, if you had to define a class method that returned an object of the class itself, it would look something like this:

```
from typing import TypeVar

T = TypeVar('T', bound=type)

class Circle:
    def __init__(self, radius: int) -> None:
        self.radius = radius

    @classmethod
    def from_diameter(cls: T, diameter) -> T:
        circle = cls(radius=diameter/2)
        return circle
```

To be able to say that a method returns the same type as the class itself, you had to define a TypeVar and say that the method returns the same type T as the current class itself.

With Self Type

Thanks to its straightforward and compact syntax, as specified in PEP 673, the "Self" type serves as the recommended annotation for methods returning an instance of their class. Starting from Python version 3.11 and onward, you can directly import the "Self" type from Python's "typing" module. For Python versions prior to 3.11, you can access the "Self" type via the "typing_extensions" module.

```python
from typing import Self

class Language:

    def __init__(self, name, version, release_date):
        self.name = name
        self.version = version
        self.release_date = release_date

    def change_version(self, version) -> Self:
        self.version = version
        return Language(self.name, self.version, self.release_date)

lang = Language("Python", 3.11, "November")
lang.change_version(3.12)
print(lang.version)
```

Output:

3.12

Improved Exceptions

Better Error Messages

Until now, in a traceback, the only information you got about where an exception got raised was the line. In Python 3.11, the exact error locations in tracebacks are showed:

```
Traceback (most recent call last):
  File "asd.py", line 15, in <module>
    print(get_margin(data))
          ^^^^^^^^^^^^^^^^
  File "asd.py", line 2, in print_margin
```

```
    margin = data['profits']['monthly'] / 10 + data['profits']['yearly'] / 2
                                                ~~~~~~~~~~~~~~~~~~~~~~~~~^~~
TypeError: unsupported operand type(s) for /: 'NoneType' and 'int'
```

Exception Notes

Python 3.11 introduces exception notes (PEP 678). Now, inside your except clauses, you can call the add_note() function and pass a custom message when you raise an error.

Example:

```
import math

try:
  math.sqrt(-1)
except ValueError as e:
  e.add_note("Negative value passed! Please try again.")
  raise
```

Another Way to Add Exception Notes: Define It As an Attribute to a Custom-Defined Exception Class

Example:

```
import math

class MyOwnError(Exception):
    __notes__ = ["This is a custom error!"]

try:
   math.sqrt(-1)
except:
   raise MyOwnError
```

Exception Groups

One way to think about exception groups (Figure 4-1) is that they're regular exceptions wrapping several other regular exceptions. However, while exception groups behave like regular exceptions in many respects, they also support special syntax that helps you handle each of the wrapped exceptions effectively. In Python 3.11, we group exceptions with ExceptionGroup().

Example:

```
def exceptionGroup():
    exec_gr = ExceptionGroup('ExceptionGroup Message!',
                [FileNotFoundError("This File is not found"),
                ValueError("Invalid Value Provided"),
                ZeroDivisionError("Trying to divide by 0")])
    raise exec_gr
```

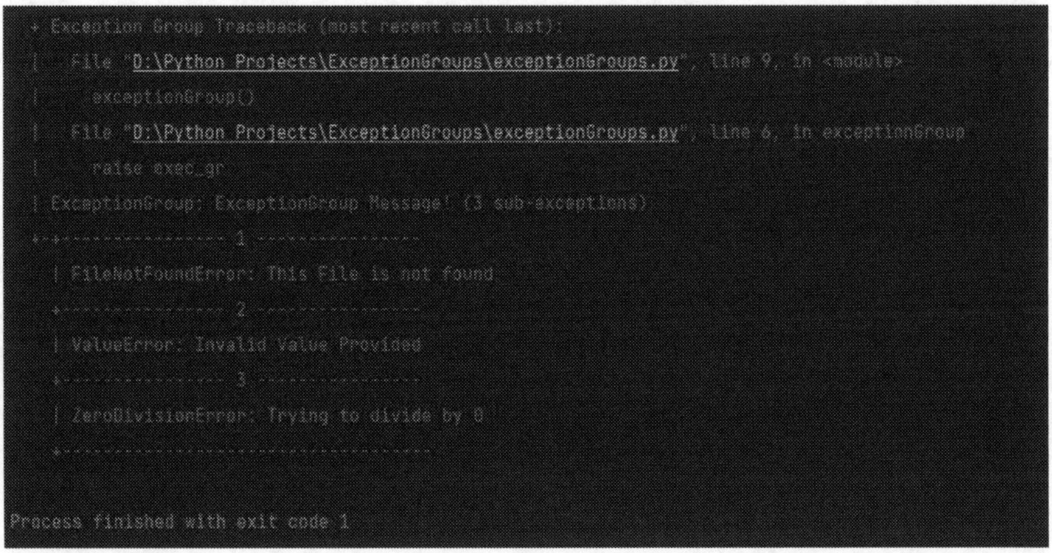

Figure 4-1. Grouped exceptions in Python

TOML Support

TOML is short for Tom's Obvious Minimal Language. It's a configuration file format that's gotten popular over the last decade. The Python community has embraced TOML as the format of choice when specifying metadata for packages and projects.

TOML has been designed to be easy for humans to read and easy for computers to parse. While TOML has been used for years by many different tools, Python hasn't had built-in TOML support. That changes in Python 3.11, when tomllib is added to the standard library. This new module builds on top of the popular tomli third-party library and allows you to parse TOML files.

Example:

```
[second]
label   = { singular = "second", plural = "seconds" }
aliases = ["s", "sec", "seconds"]

[minute]
label      = { singular = "minute", plural = "minutes" }
aliases    = ["min", "minutes"]
multiplier = 60
to_unit    = "second"

[day]
label      = { singular = "day", plural = "days" }
aliases    = ["d", "days"]
multiplier = 24
to_unit    = "hour"

[year]
label      = { singular = "year", plural = "years" }
aliases    = ["y", "yr", "years", "julian_year", "julian years"]
multiplier = 365.25
to_unit    = "day"
```

The new tomllib library brings support for parsing TOML files. tomllib does not support writing TOML. It's based on the tomli library. When using tomllib.load(), you pass in a file object that must be opened in binary mode by specifying mode="rb".

The two main functions in tomllib are as follows:

- load(): load bytes from file
- loads(): load from str

Example:

```
import tomllib

# gives TypeError, must use binary mode
with open('t.toml') as f:
    tomllib.load(f)
```

```
# correct
with open('t.toml', 'rb') as f:
    tomllib.load(f)
```

Improved Type Variables

- Arbitrary literal string type
- Negative zero formatting
- Improved type variables
- Variadic generics

Arbitrary Literal String Type

```
from typing import Literal


def paint_color(color: Literal["red", "green", "blue", "yellow"]):
    pass


paint_color("cyan")
```
> Expected type 'Literal["red", "green", "blue", "yellow"]', got 'Literal["cyan"]' instead

To address this limitation, Python 3.11 introduces a new general type LiteralString, which allows the users to enter any string literals, like the following:

```
from typing import LiteralString

def paint_color(color: LiteralString):
    pass

paint_color("cyan")
paint_color("blue")
```

Variadic Generics

```
from typing import Generic, TypeVar

Dim1 = TypeVar('Dim1')
Dim2 = TypeVar('Dim2')
Dim3 = TypeVar('Dim3')

class Shape1(Generic[Dim1]):
    pass
class Shape2(Generic[Dim1, Dim2]):
    pass
class Shape3(Generic[Dim1, Dim2, Dim3]):
    Pass
```

As shown, for three dimensions, we'll have to define three types and their respective classes, which isn't clean and represents a high level of repetition that we should be cautious about. Python 3.11 is introducing the TypeVarTuple that allows you to create generics using multiple types.

```
from typing import Generic, TypeVarTuple
Dim = TypeVarTuple('Dim')
class Shape(Generic[*Dim]):
    Pass
```

Negative Zero Formatting

Normally, there's only one zero, and it's neither positive nor negative. One weird concept that you may run into when doing calculations with floating-point numbers is negative zero.

```
small_num = -0.00321
print(f"{small_num:z.2f}")
# 0.00
```

Understanding the Role of Python 3.11 in AI and NLP – Why Python?

Why Python for AI?

Python is a popular and widely used programming language for artificial intelligence (AI) for several compelling reasons:

1. **Rich Ecosystem:** Python has a vast and robust ecosystem of libraries and frameworks specifically designed for AI and machine learning, such as TensorFlow, PyTorch, scikit-learn, Keras, and more. These libraries provide pre-built tools and functions for tasks like data manipulation, model development, and training, making AI development more accessible and efficient.

2. **Community Support:** Python has a large and active community of developers, researchers, and data scientists who contributed to the AI ecosystem. This results in constant updates, improvements, and a wealth of resources, including documentation, tutorials, and online forums for support.

3. **Easy to Learn and Read:** Python is known for its simplicity and readability. Its clear and concise syntax makes it an excellent choice for both beginners and experienced programmers. This simplicity reduces the barrier to entry for individuals looking to get started in AI.

4. **Versatility:** Python is a versatile language that can be used for a wide range of tasks beyond AI, including web development, data analysis, scripting, and more. This versatility means that you can apply Python across different domains and integrate AI solutions into various projects.

5. **Cross-Platform Compatibility:** Python is available on multiple platforms, including Windows, macOS, and various Linux distributions. This cross-platform compatibility ensures that AI projects can run seamlessly across different environments.

6. **Integration Capabilities:** Python can easily integrate with other languages and tools, making it suitable for AI development in various contexts. For example, it can interface with C/C++ libraries, Java applications, and web services, enabling you to incorporate AI into existing systems.

7. **Extensive Data Science Support:** Python is a popular choice for data scientists, and it provides powerful libraries like NumPy, pandas, and Matplotlib for data manipulation, analysis, and visualization. These tools are crucial for AI, as they help preprocess data and evaluate model performance.

8. **Open Source:** Python and most of its AI libraries are open source, which means they are freely available and can be customized or extended as needed. This open nature fosters innovation and collaboration in the AI community.

9. **Industry Adoption:** Python has gained widespread adoption in the AI and machine learning industry. Many tech companies and research institutions use Python as their primary language for AI development, contributing to its popularity and growth.

Why Python for NLP?

Python is a preferred language for natural language processing (NLP) for several compelling reasons:

1. **Rich NLP Libraries:** Python boasts a wide range of NLP libraries and frameworks, such as NLTK (Natural Language Toolkit), spaCy, Gensim, TextBlob, and more. These libraries provide pre-built tools and resources for tasks like text tokenization, part-of-speech tagging, sentiment analysis, named entity recognition, and language modeling, simplifying NLP development.

2. **Active NLP Community:** Python has a vibrant NLP community of developers, researchers, and linguists who continually contribute to and improve NLP libraries and resources. This results in frequent updates, new models, and a wealth of tutorials and documentation to assist NLP practitioners.

3. **Machine Learning Integration:** NLP often involves machine learning techniques, and Python's extensive machine learning libraries, such as scikit-learn, TensorFlow, and PyTorch, make it convenient to build and train NLP models.

4. **Pre-trained Models:** Python-based NLP libraries often provide pre-trained models for various NLP tasks. For example, spaCy and Hugging Face Transformers, OpenAI, and LangChain offer pre-trained models for tasks like text classification, translation, and question answering, saving developers time and effort.

5. **Open Source Ecosystem:** Most NLP libraries in Python are open source, allowing developers to access, modify, and extend them as needed. This fosters collaboration and innovation within the NLP community.

6. **Wide Adoption:** Python is widely adopted in academia, research, and industry for NLP-related projects. Many leading research papers, conferences, and organizations use Python as their primary language for NLP research and development.

Summary

Python's syntax is designed for readability and simplicity, adhering to the philosophy that there should be a clear way to accomplish tasks. This design choice, coupled with English keywords instead of punctuation, makes Python's code clean and easy to understand.

In the next chapter, we will discuss the following:

- The different types of layers in the architecture of the large language models
- Attention mechanisms
- Tokenization strategies in large language models
- What is zero-shot and few-shot learning
- More about LLM hallucinations
- Examples of architectures of popular LLMs

CHAPTER 5

Basic Overview of the Components of the LLM Architectures

This chapter delves into the intricate components that constitute large language model (LLM) architectures. Understanding these elements is crucial for appreciating how LLMs transform raw textual data into meaningful, context-aware outputs. The key components discussed in this chapter include **embedding layers, feedforward layers, recurrent layers, and attention mechanisms**. Each of these plays a pivotal role in enabling LLMs to process and generate human-like language.

Embedding layers are foundational to LLMs, converting discrete tokens such as words into continuous vector representations. The embedding matrix within these layers, initially filled with random values, is a learnable component that adapts during training. Each row of this matrix corresponds to a unique token from the model's vocabulary. The embedding layer fetches these rows to create vector representations, which serve as the model's understanding of the tokens' meanings. Various embedding algorithms enhance the model's ability to capture positional and contextual nuances, enabling it to handle diverse linguistic scenarios effectively.

Feedforward neural networks (FFNs) are a straightforward type of artificial neural network, characterized by the unidirectional flow of data from input to output nodes through hidden layers. Each neuron in these layers is connected to neurons in adjacent layers via weighted connections. FFNs process input data by calculating weighted sums, applying activation functions, and generating outputs. In LLMs, feedforward layers are integral to recognizing and assimilating complex patterns, thereby enhancing the model's comprehension capabilities.

Recurrent layers are crucial for processing sequential data, particularly in natural language processing (NLP). These layers maintain hidden states that carry forward information from previous elements in a sequence, allowing the model to consider the context of earlier words when interpreting current ones. Training recurrent layers involves techniques like backpropagation through time (BPTT), which adjusts weights based on the model's performance across entire sequences. Variants like Long Short-Term Memory (LSTM) units and Gated Recurrent Units (GRUs) are employed to manage long-term dependencies more effectively.

Attention mechanisms are a cornerstone of modern LLMs, enabling them to focus on specific parts of the input text that are most relevant to the task at hand. The attention mechanism functions through a two-phase approach: during the attention phase, words seek out and exchange information with relevant words in the context, and during the feedforward phase, they process this accumulated information. Transformers, a type of LLM, utilize multiple attention heads to perform various information-exchange tasks in parallel, enhancing their ability to process extensive texts efficiently.

Self-attention allows each word in a sentence to attend to all other words, capturing their contextual relationships. Multi-head attention extends this by enabling the model to focus on different parts of the sentence simultaneously, thereby improving its understanding of syntactic and semantic relationships. These mechanisms are vital for tasks requiring a comprehensive understanding of text, such as translation and summarization.

Tokenization is the process of breaking down text into manageable units known as tokens, which are then transformed into numerical representations called embeddings. This segmentation is fundamental for LLMs to process and learn from large datasets efficiently. The distribution and handling of these tokens, especially predicting the next token in a sequence, form the basis of many LLM tasks, including text generation and translation.

Embedding Layers

Embedding layers in neural networks, including large language models (LLMs), play a crucial role in transforming discrete inputs, such as words or tokens, into continuous vector forms. At the heart of this transformation is the **embedding matrix**, a learnable component of the layer that starts off with random values.

CHAPTER 5   BASIC OVERVIEW OF THE COMPONENTS OF THE LLM ARCHITECTURES

This matrix features rows that correspond to the unique tokens in the model's vocabulary. In operation, the embedding layer conducts a lookup for each input token, fetching the corresponding row from the matrix. These fetched rows serve as the **continuous vector representations, or embeddings (Figure 5-1), of the tokens**. This outlines the fundamental operation of embedding layers, although each embedding algorithm may vary, incorporating the position and context of sentences. Such variations enable different models to perform with varying effectiveness across diverse scenarios.

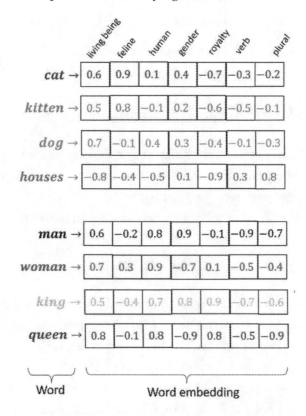

Figure 5-1. Word embeddings visualization (Source: http://medium.com)

Word embeddings (Figure 5-2) represent individual words within a text as vectors, generated by specialized models designed for this purpose. Each word is assigned a distinct vector, which is essentially an array filled with numbers that uniquely identify that word. These vectors are multidimensional entities, with each dimension representing a numerical component specific to the word. The uniqueness of these vectors allows for the distinct representation of each word within a document.

201

The principle of word embedding is to map words with similar meanings close to each other in the vector space. For instance, the vector assigned to "apple" would be more similar to that of "orange" than to "violin," reflecting the closer relationship between the fruits compared to the musical instrument. This similarity in vectors mirrors the semantic proximity between words.

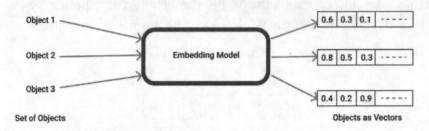

Figure 5-2. Embeddings (Source: http://medium.com)

It's important to mention that the choice of dimensionality for these vectors lies in the hands of the model's architect. Take **Word2Vec** as an instance; it employs vectors of 300 dimensions, representing each word with 300 distinct numbers. In this context, we aim to construct a basic word embedding model using just two dimensions.

Stage 1: Nodes

The initial phase involves establishing a series of nodes grouped within a "hidden layer" (Figure 5-3). Each node is linked to an input word through a weighted connection. Nodes are structured in two segments. The first segment aggregates the weighted inputs from each word. This aggregated sum is then forwarded to the second segment, the activation function. The role of the activation function is to determine the node's output based on its specific process. In this scenario, the activation functions act as simple identity functions, leaving the input unchanged.

The total number of nodes determines the vector dimensionality, as previously discussed. This means it sets the quantity of numbers (or vector components) assigned to each input word. Typically, in practical applications, the count of nodes ranges from tens to hundreds.

Every word in the dataset is fed into each node with a specific weight, aiming to facilitate the model's learning of word relationships without increasing the system's complexity.

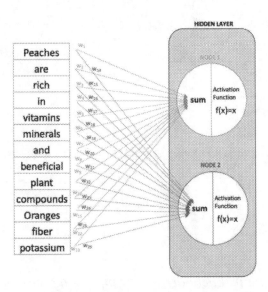

Figure 5-3. *Input words are processed through a hidden layer in a neural network (Source: http://medium.com)*

In the initial stage, as outlined, each word is linked to nodes through weighted connections, represented as (w_1, w_2, ... w_n). These weights are initially set to random values by the models to kick-start the learning process. The objective is to adjust and refine these weights through numerous iterations to enhance model accuracy. Naturally, it's not feasible to perform mathematical operations like multiplication between a string (word) and a number directly. Hence, a binary approach is employed where words are encoded as 0 or 1. Specifically, the word immediately preceding the target word for prediction is assigned a value of 1.

Stage 2: Returning to the Words

During this second stage, the outputs generated by each node are allocated back to the words (Figure 5-4), utilizing weights that have been assigned randomly.

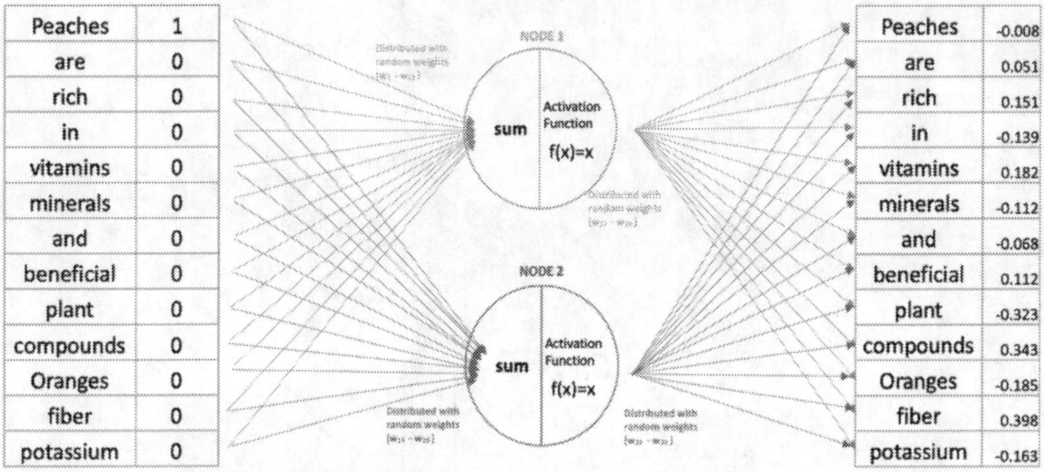

Figure 5-4. Detailed neural network diagram with a focus on the hidden layer and includes specific weights for connections

Stage 3: Implementing the Softmax Layer

In this third stage, we incorporate a softmax layer into our model. This layer functions as an activation mechanism, predominantly used in the concluding output layer for multiclass classification scenarios. The purpose of the softmax layer (Figure 5-5) is to convert a series of numerical values into a distribution of probabilities. It processes a numerical value associated with each category (or word, in our context), performs calculations, and generates a probability score for each category. Unlike previous layers that might use randomly assigned weights, the softmax layer employs a specific formula for its operations.

This function processes an input vector, applying its formula to each element within the vector. For instance, given an input vector like [3, 5, 7], the softmax layer would output a vector such as [0.016, 0.117, 0.867], where the sum of the output vector's elements equals 1, indicating a probability distribution.

$$x = \begin{bmatrix} 3 \\ 5 \\ 7 \end{bmatrix} \qquad softmax(x) = \begin{bmatrix} \frac{e^3}{e^3+e^5+e^7} \\ \frac{e^5}{e^3+e^5+e^7} \\ \frac{e^7}{e^3+e^5+e^7} \end{bmatrix} = \begin{bmatrix} 0.016 \\ 0.117 \\ 0.867 \end{bmatrix}$$

Figure 5-5. Softmax function calculation

Feedforward Layers

Feedforward layers in a neural network are layers where the connections between nodes do not form cycles. Each node in one layer connects only to nodes in the next layer, allowing information to flow in one direction – from input to output. This structure is fundamental in neural networks for tasks like classification and regression.

What Is a Feedforward Neural Network?

A **feedforward neural network** (or, as they are often called, "FFNs") stands as one of the simplest types of artificial neural networks, characterized by the unidirectional flow of data from input to output nodes, traversing through hidden nodes without any cycles or loops. This straightforward design distinguishes it from more complex networks such as recurrent neural networks (RNNs) and convolutional neural networks (CNNs), due to its absence of feedback loops, making it easier to build.

Its architecture is structured around three main components: **input layers, hidden layers, and output layers**. Each layer is composed of units known as neurons, with each neuron in one layer fully connected to neurons in the subsequent layer through weighted connections.

In operation, neurons in the input layer receive data and pass it forward. The number of neurons in the input layer matches the size of the input data. The hidden layers, situated between the input and output, perform computations while remaining hidden from both the input and output.

Neurons within these layers calculate a weighted sum from the preceding layer's outputs, apply an activation function to introduce nonlinearity, and forward the result onward. This process is iterative until the output layer is reached, which then generates the network's final output. The connectivity between neurons across layers renders feedforward neural networks fully connected, with weights symbolizing the connection strength that the network adjusts during learning to minimize output errors.

The operation of feedforward neural networks unfolds in two critical phases: **the feedforward and the backpropagation phases**.

Feedforward Phase

During this phase, the network processes the input by advancing it through the layers. Hidden layers calculate the weighted sum of inputs and then apply an activation function (e.g., ReLU, Sigmoid, TanH) to incorporate nonlinearity, continually forwarding the data until the output layer is reached and a prediction is made.

Backpropagation Phase

Following a prediction, the network evaluates the discrepancy between the actual output and the expected output. This error is then propagated backward through the network. To reduce the error, the network employs a **gradient descent optimization technique**, adjusting the weights across connections accordingly.

LLMs and Feedforward Layers

Large language models include feedforward layers, which are composed of multiple interconnected layers that execute nonlinear transformations on the input data. This structure helps the model to recognize and assimilate **complex patterns hidden in the input text**, enhancing its comprehension of the material. These layers assist the model in interpreting the user's intended message in the text by extracting more complex concepts.

In a **feedforward network (Figure 5-6)**, the system processes each word vector independently, aiming to forecast the subsequent word without any interaction between words at this point. The feedforward layer examines each word separately but can utilize any data previously captured by an attention mechanism. The architecture of the feedforward layer within GPT-3's most extensive model is designed to function in this manner.

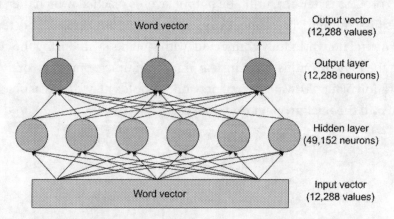

Figure 5-6. *Feedforward neural network*

The colored circles represent neurons, which are essentially mathematical functions calculating a weighted sum of their inputs. The strength of the feedforward layer lies in its extensive network of connections. Although illustrated with a modest number of neurons, the actual scale in GPT-3's feedforward layers is vastly larger, featuring 12,288 neurons in the output layer to match its 12,288-dimensional word vectors, and a staggering 49,152 neurons in the hidden layer.

*In GPT-3's most comprehensive version, the hidden layer alone comprises **49,152 neurons**, each receiving 12,288 inputs, equating to the same number of weight parameters per neuron. The output layer contains **12,288 neurons**, each with **49,152 inputs**, resulting in **49,152 weight parameters per neuron**. This configuration results in over **1.2 billion** weight parameters per feedforward layer, and with 96 such layers, the total parameter count reaches **116 billion**. This massive number represents nearly two-thirds of GPT-3's total parameter count of **175 billion**.*

Research by Tel Aviv University in 2020[1] revealed that feedforward layers function through pattern recognition: each neuron in the hidden layer identifies specific patterns within the input text. For instance, in a 16-layer GPT-2 model, neurons were found to recognize patterns ranging from word sequences ending with "substitutes" to those pertaining to military bases, time intervals, and television shows, showcasing an increase in the abstraction level of recognized patterns across layers. Early layers pinpointed specific words, while deeper layers identified phrases within broader semantic fields.

This insight is particularly fascinating given that the feedforward layer processes each word individually. For example, when identifying a phrase as television related, it relies on the vector for "archived" without direct access to associated words like "NBC" or "daytime". This suggests that attention mechanisms preceding the feedforward layer integrate contextual information into the vector, enabling pattern recognition.

Upon recognizing a pattern, a neuron enriches the word vector with additional information, which, while sometimes abstract, can often hint at a probable following word.

---

[1] Mor Geva, Roei Schuster, Jonathan Berant, Omer Levy, Transformer Feed-Forward Layers Are Key-Value Memories

CHAPTER 5   BASIC OVERVIEW OF THE COMPONENTS OF THE LLM ARCHITECTURES

Recurrent Layers

The last layer of large language models (LLMs) is known as the recurrent layer. This layer is responsible for understanding the words in the input text sequence, enabling it to grasp the connections among various words within the sequence presented in user prompts.

Recurrent layers in large language models (LLMs) are pivotal for processing sequential data, particularly in the domain of natural language processing (NLP). These layers enable LLMs to handle temporal dependencies and contextual nuances in text data, leveraging the sequential nature of language.

Here's a Closer Look at How Recurrent Layers Function Within LLMs

Sequential Data Processing

Unlike feedforward layers that process inputs independently, recurrent layers are adept at handling sequential data. They achieve this by looping through each element of the input sequence, carrying forward a state that contains information about previously seen elements. This state acts as a memory that influences the processing of future elements, allowing the model to consider the context of earlier words when interpreting the current word.

Hidden States

At the core of a recurrent layer is the hidden state, which is updated at each step of the sequence. When a new input is received, the recurrent layer combines it with the existing hidden state to produce a new hidden state. This mechanism enables the model to accumulate knowledge about the entire sequence up to the current point, which is essential for understanding the structure and meaning of the text.

Backpropagation Through Time

Training recurrent layers involves a technique known as backpropagation through time (BPTT). This method unfolds the recurrent network across each time step, treating it as a deep feedforward network where each layer corresponds to a time step in the sequence. BPTT allows the model to learn long-term dependencies by adjusting weights based on the gradient of the loss function, which considers the performance of the model across the entire sequence.

Challenges and Solutions

One challenge with recurrent layers is the difficulty in learning long-term dependencies due to issues like vanishing and exploding gradients. LLMs often incorporate advanced variants of recurrent layers, such as Long Short-Term Memory (LSTM) units or Gated Recurrent Units (GRUs), to mitigate these problems. These variants introduce gates that regulate the flow of information, making it easier to remember or forget certain information, thereby enhancing the model's ability to learn from long sequences.

In the context of LLMs, recurrent layers contribute to tasks like text generation, translation, and sentiment analysis by providing a nuanced understanding of the sequence's context. They enable the model to generate coherent and contextually relevant text by remembering what has been said previously and using that information to influence what comes next.

The components of the architecture of LLMs also emphasize the significance of the attention mechanism. Large language models (LLMs) employ this mechanism to concentrate on specific segments of the input text that are pertinent to the task at hand. The inclusion of a self-attention mechanism layer aids the model in producing outputs of higher precision.

Attention Mechanisms

Let's explore the internal workings of a transformer in processing input text. The mechanism involves a dual-phase approach to refresh the contextual information for every word:

- Initially, during the **attention phase**, words engage in seeking out others within the context that are of relevance, allowing for an exchange of information.

- Subsequently, in the **feedforward phase**, each word processes the accumulated insights from the attention phase to anticipate the upcoming word.

It's crucial to note that these operations are conducted by the network as a whole rather than by individual words. This clarification is important to highlight the transformer's method of analyzing text at the word level, as opposed to larger text blocks. This strategy leverages the extensive parallel processing capabilities of contemporary GPUs, enabling large language models (LLMs) to efficiently process extensive texts, an aspect where previous models faced limitations.

The attention mechanism functions akin to a matchmaking system for words, **where each word generates a query vector outlining the traits of words it seeks and a key vector detailing its own traits**. The network evaluates the compatibility of key and query vectors through dot product calculations to identify matching words, facilitating the exchange of information between them.

Consider a scenario where a transformer determines that "his" refers to "John" in a sentence fragment. This process involves matching the query vector for "his," which might be looking for a noun identifying a male person, with the key vector for "John," indicating it as a noun for a male person. Upon finding a match, the network transfers information from "John" to "his."

Transformers feature multiple "attention heads" in each attention layer, enabling parallel processing of various information-exchange tasks. These tasks range from associating pronouns with nouns, clarifying homonyms, to linking multi-word phrases. The operation of attention heads is sequential, with the output from one serving as the input for another in the next layer, often requiring multiple attention heads for complex tasks.

In the case of GPT-3's largest variant, it consists of 96 layers, each with 96 attention heads, culminating in 9,216 attention operations for every word prediction.

Large language models (LLMs), such as those based on the transformer architecture, utilize sophisticated attention mechanisms to process and generate language. These mechanisms are pivotal for understanding the context and relationships between words in a sentence. Let's explore the primary types of attention mechanisms used in LLMs.

Self-attention (Intra-attention)

Self-attention, a cornerstone of the transformer model, allows each word in a sentence to attend to all other words to capture their contextual relationships. This mechanism helps the model understand the meaning of each word within the context of the entire sentence. It's particularly effective in identifying dependencies and relationships, regardless of their distance in the text.

Multi-head Attention

Multi-head attention is an extension of self-attention that allows the model to focus on different parts of the sentence simultaneously. By dividing the attention mechanism into multiple "heads," the model can capture various aspects of word context, such as syntactic and semantic relationships, in parallel. This leads to a more comprehensive understanding of the text.

Cross-Attention (Encoder-Decoder Attention)

Cross-attention is used in models with encoder-decoder structures, where the decoder attends to the encoder's output. This mechanism is crucial for tasks like translation, where the model needs to consider the input sequence (encoded information) while generating the output sequence. Cross-attention helps the decoder focus on relevant parts of the input text, improving the accuracy of the generated output.

Masked Attention

Used primarily in the decoder part of the transformer, masked attention prevents positions from attending to subsequent positions. This is essential during training to ensure that the prediction for a particular word only depends on previously generated words, maintaining the autoregressive property. Masked attention is key in generating coherent and contextually appropriate text.

Sparse Attention

Sparse attention mechanisms are designed to improve efficiency and scalability for processing long sequences of text. By selectively focusing on a subset of the input positions, sparse attention reduces computational complexity. Models like Longformer and BigBird implement variations of sparse attention to handle longer documents effectively.

Global/Local Attention

Some models incorporate global and local attention mechanisms to balance between focusing on the entire text and concentrating on specific, relevant segments. Global attention may consider all input tokens, while local attention focuses on a neighborhood around the current token, optimizing both performance and computational efficiency.

These attention mechanisms enable LLMs to process and understand language at an unprecedented scale, handling complex linguistic patterns and nuances. By leveraging these diverse attention strategies, LLMs achieve remarkable performance across a wide range of natural language processing tasks, from translation and summarization to question answering and creative writing.

CHAPTER 5   BASIC OVERVIEW OF THE COMPONENTS OF THE LLM ARCHITECTURES

Understanding Tokens and Token Distributions and Predicting the Next Token

In NLP, tokens are the smallest units of text, like words or characters. Tokenization breaks text into these pieces for analysis. Token distribution refers to the frequency and patterns of tokens in a text. Understanding this helps in model training by focusing on common patterns. Predicting the next token uses models like n-grams, hidden Markov models, and neural networks to analyze the context of previous tokens. This enhances applications such as text generation, autocomplete, and language translation.

Understanding Tokenization in the Context of Large Language Models

Large language models (LLMs) like GPT-3 revolutionize text processing by dissecting vast textual data into manageable units, known as tokens. This segmentation into tokens serves as a crucial step, enabling these models to digest and learn from extensive datasets efficiently.

Imagine tokenization as a **master key unlocking the potential of LLMs** to assimilate and interpret massive volumes of text. By converting text into tokens, these models can effortlessly navigate through and process large datasets. Tokenization acts as a precision tool, segmenting text into digestible bits, thereby allowing the models to scale their understanding and processing capabilities. It ensures that no dataset is too voluminous or complex, making it foundational for training models on expansive text collections.

Tokenization serves as a pivotal mechanism, especially in the realm of natural language processing (NLP), by enabling LLMs to dissect and utilize even the most daunting text compilations. This process not only enhances the models' comprehension and generation capabilities but also significantly broadens their scope of knowledge and application.

The Advantages of Tokenization for LLMs

Tokenizing text into smaller elements yields multiple advantages for LLMs, enhancing their efficiency and effectiveness in language understanding:

- **Improved Data Management:** Tokenization simplifies data handling, allowing models to navigate and interpret linguistic trends and patterns more accurately.

- **Resource Optimization:** It facilitates more efficient use of memory and computational resources, paving the way for more robust and scalable models.

- **Versatility in Applications:** Standardizing input through tokenization ensures LLMs can be seamlessly integrated across diverse fields, from healthcare to finance, enhancing interoperability among various text analysis systems.

Limitations and Challenges

Despite its critical role, tokenization is not without its challenges. It can sometimes lead to the loss of nuanced information, particularly with languages or texts rich in context, subtleties, or unconventional formats. Moreover, tokenization may struggle with words outside its predefined vocabulary or languages that are character based, presenting obstacles in fully capturing the intricacies of human language.

In essence, while tokenization is a cornerstone in the development and functionality of large language models, navigating its limitations remains a key area of focus to unlock even greater capabilities in language understanding and processing.

Challenges in Current Tokenization Techniques

Tokenization techniques face several challenges. These include handling diverse languages and dialects, managing out-of-vocabulary words, dealing with complex morphology, and addressing context sensitivity. Additionally, tokenization can struggle with idiomatic expressions and varying token lengths, impacting the performance of language models. Overcoming these challenges is crucial for improving the accuracy and efficiency of NLP applications.

Tokenization, a critical step in processing text for natural language models, faces several notable challenges.

Case Sensitivity in Tokenization

Tokenizers often differentiate between the same words based on their case. For instance, "hello" might be assigned a single token ID (e.g., 31373), whereas "HELLO" could be broken down into multiple tokens such as [13909, 3069, 46], corresponding to the segments ["HE", "LL", "O"]. This variance introduces complexity in handling case-sensitive words.

Numeric Data Handling

Transformers, by design, show limited aptitude for numerical tasks, partly due to the tokenizer's erratic representation of numbers. A number like 200 might be encapsulated as a singular token, whereas 201 could be split into multiple tokens (e.g., [20, 1]), leading to inconsistent handling of numerical data.

Inconsistencies with Trailing Whitespace

Tokenizers can exhibit unpredictable behavior with trailing whitespaces. A word such as "last" might be tokenized with an appended whitespace (" last"), differing from the expected tokenization into separate entities for the space and the word. This inconsistency can affect the model's ability to accurately predict subsequent words, depending on whether the input ends with a whitespace.

Model-Specific Tokenization Practices

Despite the widespread adoption of the byte pair encoding (BPE) method or similar approaches in tokenization, language models like GPT-4, LLaMA, and OpenAssistant each develop bespoke tokenizers. This means that even models employing the same foundational tokenization strategy may have unique implementations, affecting interoperability and consistency across different systems.

Grasping Contextual Nuances

The division of text into individual tokens can lead to a dilution of context, a challenge particularly pronounced in languages rich with nuanced expressions or ideas heavily reliant on context. This potential loss of context can compromise the effectiveness of natural language understanding systems.

Navigating Ambiguity

Tokenization may struggle to address lexical ambiguity, where a single word carries multiple meanings. Deciphering the correct interpretation can prove to be a complex puzzle.

Interpreting Idioms

The segmentation of idiomatic expressions into discrete tokens risks stripping them of their inherent meanings, as these phrases derive their significance from the specific combination of words used together.

Handling Special Symbols and Characters

The presence of unique symbols, punctuation, or characters poses challenges for tokenization strategies, potentially hindering their ability to accurately process and interpret text, especially in fields that employ technical language or specialized terminology.

Tokenization Strategies in Large Language Models

Tokenization plays an indispensable role in the preprocessing phase of natural language processing (NLP), employing a range of techniques from simple space-based segmentation to more sophisticated methods such as fragment splitting and binary code pairing. The choice of tokenization technique is influenced by the specific requirements of the NLP task, the characteristics of the language, and the nature of the dataset involved.

Common Tokenization Methods

- **Word Tokenization:** This method segments text into individual words and stands as the most frequently used tokenization technique.
- **Sentence Tokenization:** This approach divides text into its constituent sentences, facilitating tasks that require understanding of sentence boundaries.

- **Subword Tokenization:** Subword tokenization breaks down words into smaller components, addressing the challenge of morphological diversity within languages.

- **Character Tokenization:** By splitting text down to its individual characters, this method allows for a granular analysis of the text data.

- **Byte Pair Encoding (BPE):** BPE operates through an iterative process, merging the most commonly occurring pairs of consecutive tokens to build a comprehensive vocabulary. This algorithm is particularly effective in managing the vast and varied vocabularies inherent in large language models, striking a balance between the granularity of character tokenization and the efficiency of word tokenization.

These tokenization techniques provide the foundational tools for large language models to process, understand, and generate human language with remarkable accuracy and nuance.

What Is Token Distribution?

The token distribution in large language models (LLMs), such as GPT (Generative Pre-trained Transformer) models, refers to how the model processes and generates text based on a sequence of tokens. In the context of these models, "tokens" are the basic units of text that the model understands and manipulates. The concept of token distribution can be understood from several perspectives:

- **Tokenization Process:** Text input is converted into tokens using a tokenization process. This process can vary; for example, it might involve splitting the text into words, subwords, or even characters, depending on the tokenizer's design. The choice of tokenization impacts the model's vocabulary size and its ability to handle different languages or specialized jargon.

- **Vocabulary Distribution:** The vocabulary of a large language model consists of a fixed set of tokens that the model has been trained on. This vocabulary is designed to cover as wide a range of text as possible. However, the distribution of tokens in the model's training data will influence its performance. Commonly occurring tokens or phrases in the training dataset will be better represented and understood by the model compared to rare or unseen tokens.

- **Token Frequency and Representation:** In training datasets, some tokens appear more frequently than others, leading to a skewed distribution where a small number of tokens account for a large portion of the text. Models are generally trained to handle this imbalance, but their effectiveness can vary. High-frequency tokens are often represented with greater accuracy and detail.

- **Embedding Space Distribution:** Each token in a language model's vocabulary is associated with an embedding, a vector that represents the token in a high-dimensional space. The distribution of these embeddings reflects how the model perceives relationships between tokens, including semantic and syntactic similarities.

- **Output Token Distribution:** When generating text, the model predicts the next token in a sequence based on a probability distribution over its vocabulary. This distribution is influenced by the input context, the model's internal representations, and its training. The diversity and plausibility of generated text are directly tied to how well the model manages this distribution.

- **Handling of Rare or Unseen Tokens:** Large language models employ strategies like byte pair encoding (BPE) or similar tokenization methods to deal with rare or unseen words by breaking them down into subword units. This allows the model to handle a wide variety of text inputs, even if they contain words not explicitly present in the training vocabulary.

In essence, the distribution of tokens in large language models is a critical aspect that affects their understanding, generation, and overall handling of language. It encompasses the way text is tokenized, how tokens are represented and managed within the model, and how these tokens influence the model's output.

CHAPTER 5   BASIC OVERVIEW OF THE COMPONENTS OF THE LLM ARCHITECTURES

Predicting the Next Token[2]

Large language models have made significant strides in a variety of natural language processing (NLP) tasks, including machine translation, logical reasoning, coding, and understanding natural language. Models such as GPT-3, GPT-4, and LaMDA are built upon vast text datasets, enabling them to produce responses that are both coherent and contextually appropriate to given prompts. Remarkably, these models primarily operate on a straightforward principle: **predicting the subsequent token in a sequence**.

Despite the simplicity of this approach, models trained on sufficiently comprehensive datasets have demonstrated the capability to tackle complex problems. Let's dive into the processes involved in the "next token" prediction.

1. Initially, words are segmented into tokens, which are subsequently transformed into numerical representations known as embeddings. These embeddings (Figure 5-7) maintain the semantic essence of the original words within a vector of dimensions [512, 1].

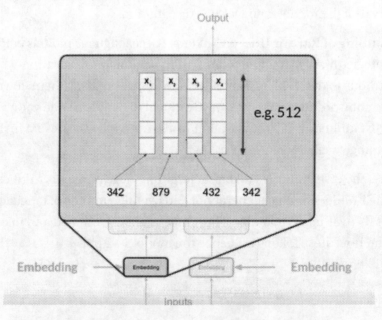

Figure 5-7. *Token embedding visualization (Source: Coursera)*

---

[2] Coursera, Natural Language Processing with Attention Models

CHAPTER 5 BASIC OVERVIEW OF THE COMPONENTS OF THE LLM ARCHITECTURES

2. When incorporating these token vectors into the encoder and decoder framework, positional encodings are also added. This technique allows the model to process all input tokens simultaneously while retaining the sequence order of the words through the use of positional information (Figure 5-8).

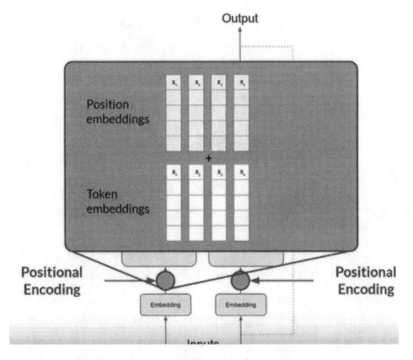

Figure 5-8. *Token and positional embeddings (Source: Coursera)*

3. The combined vectors of input tokens and positional encodings are then directed to a self-attention mechanism. Within this layer, the model evaluates the interrelationships among all tokens in the input. The self-attention mechanism is designed to learn and store weights during training, which signify the relevance of each token to every other token in the input. Notably, the transformer model employs a multi-headed self-attention approach (Figure 5-9), enabling the parallel learning of multiple self-attention weight sets, each deciphering different linguistic features.

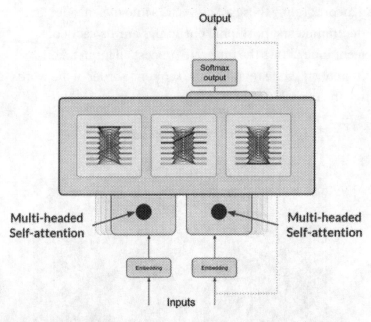

Figure 5-9. *Multi-headed self-attention mechanism (Source: Coursera)*

4. Following the application of the attention weights to the input, the resultant data is conveyed through a dense feedforward network.

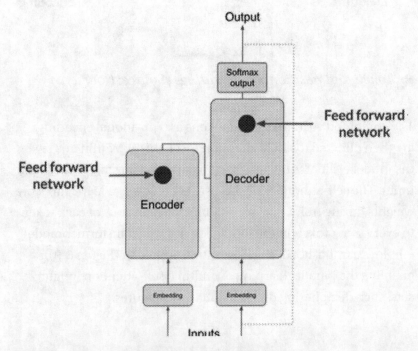

Figure 5-10. *Encoder-decoder feedforward network (Source: Coursera)*

CHAPTER 5 BASIC OVERVIEW OF THE COMPONENTS OF THE LLM ARCHITECTURES

5. This network (Figure 5-10) produces a vector of logits that correlates with the likelihood of each token within the model's vocabulary. These logits are subsequently normalized into probability scores for each vocabulary word through a softmax layer.

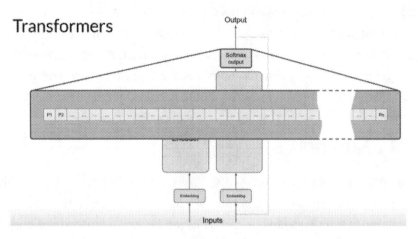

Figure 5-11. *Transformer model architecture (Source: Coursera)*

6. The transformer (Figure 5-11 and Figure 5-12) selects the word with the highest probability score, which is then fed into the decoder to facilitate the generation of the subsequent word. This iterative process continues until the model produces an end token, illustrating the procedure of language translation performed by a transformer.

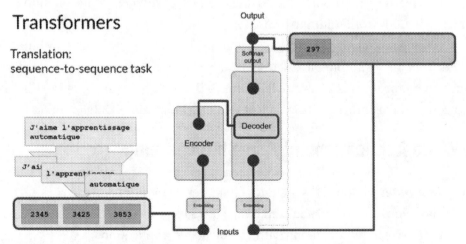

Figure 5-12. *Transformer-based sequence-to-sequence translation (Source: Coursera)*

Zero-Shot and Few-Shot Learning

Zero-shot and few-shot learning are techniques in deep learning that enable models to recognize and classify new, unseen data with little to no training examples. Zero-shot learning allows a model to make predictions based on semantic relationships and prior knowledge without any training examples for the new class. Few-shot learning, on the other hand, involves training models with a very small number of examples, leveraging transfer learning and meta-learning to generalize from limited data. These approaches are particularly useful in scenarios where labeled data is scarce.

Few-Shot Learning

Few-shot learning is a technique that involves training models on a significantly smaller dataset than is typically used. This approach is a prime example of meta-learning, where the model undergoes training across a variety of related tasks during its meta-training stage. This process equips the model to perform effectively on new, unseen data by leveraging a very limited number of examples.

The Significance of Few-Shot Learning

Few-shot learning stands out for its ability to diminish the need for extensive data collection, thus lowering both the costs associated with gathering data and the computational expenses. It shines in scenarios where traditional supervised or unsupervised learning methods struggle due to a lack of sufficient data, enabling accurate predictions from minimal examples.

This learning model mirrors human capability, such as recognizing new handwritten characters from a few samples, a task that usually requires vast amounts of data for machines. In medical fields, few-shot learning facilitates the diagnosis of rare diseases through computer vision, analyzing abnormalities with minimal data input.

Real-World Applications of Few-Shot Learning

- **Computer Vision:** This includes character recognition, image classification, and specialized image tasks like image retrieval and gesture recognition, along with video-related applications.

- **Natural Language Processing (NLP):** Few-shot learning aids in parsing, translation, sentiment analysis from concise reviews, user intent recognition, and various text classification tasks.

- **Robotics:** The technology is applied in visual navigation, executing continuous control tasks, and learning manipulation techniques from limited demonstrations.

- **Audio Processing:** It is used for converting voices across different languages and adapting a voice from one individual to another.

- **Other Fields:** Few-shot learning finds its use in a broad range of applications including medical diagnostics, IoT (Internet of Things), mathematical modeling, and material science.

Zero-Shot Learning

Zero-shot learning is an advanced approach in machine learning where a model, once trained, is capable of making predictions for classes it has never seen during its training phase. This technique draws inspiration from human cognitive abilities to recognize and relate new information based on learned concepts, enabling machines to similarly identify new categories.

The primary goal of zero-shot learning is to empower models to accurately classify or recognize objects from entirely new categories without the need for direct training on those specific classes. This capability is achieved through the transfer of knowledge from previously learned data, emphasizing the model's ability to understand and apply semantic relationships and attributes to new, unseen data.

Zero-shot learning focuses on the development of models that can interpret and utilize intermediate semantic features to identify novel classes. An illustrative example of this can be understood through the analogy of distinguishing a zebra from a horse. Even if one has never encountered a zebra, understanding that it resembles a horse with black and white stripes enables recognition upon first sight.

Significance and Use Cases of Zero-Shot Learning

- **Efficiency in Data Utilization:** Zero-shot learning significantly reduces the need for extensive data labeling, a process that is often time-consuming and resource-intensive. This approach is particularly valuable when there is a scarcity of labeled data for specific classes.

- **Adaptability:** It allows models to adapt to new tasks without the necessity to undergo retraining on previously acquired knowledge, showcasing remarkable flexibility in learning.

- **Enhanced Model Generalization:** Zero-shot learning aims to improve the generalization capabilities of models, making them more adept at handling a wide variety of tasks and data types.

- **Innovative Learning Strategy:** This method proposes an alternative to conventional learning strategies, potentially bypassing the limitations of trial-and-error learning by leveraging direct knowledge transfer.

- **Visual Recognition:** In the realms of image classification and object detection, zero-shot learning proves invaluable in identifying and classifying images or objects without prior exposure.

- **Deep Learning Applications:** Zero-shot learning is instrumental in advancing various deep learning frameworks, including image generation and retrieval, by enabling models to generate or find images based on descriptions of unseen objects.

Through these applications, zero-shot learning presents a transformative approach in machine learning, offering a bridge to more intuitive, efficient, and versatile models capable of understanding and acting upon the world in ways that mirror human learning.

Navigating Limited Data Learning: Few-Shot, One-Shot, and Zero-Shot Learning Explained

In the realm of machine learning, training models with scarce data presents a unique set of challenges and solutions. Among these, few-shot, one-shot, and zero-shot learning stand out as innovative approaches designed to leverage minimal data in distinct ways.

- **Few-Shot Learning** is tailored for situations where only a limited dataset is available for training. This approach enables models to learn and perform tasks such as image classification and facial recognition by training on just a small collection of data samples. Few-shot learning is particularly useful when the data is scarce but still available for model training.

- **One-Shot Learning** further minimizes the data requirement by training models using a single data instance or example. This method drastically reduces the need for extensive datasets, making it highly effective for specific recognition tasks, such as verifying an individual's identity from a single piece of identification. One-shot learning showcases the ability of models to generalize from a single example to understand and recognize new instances.

- **Zero-Shot Learning**, on the other hand, operates under conditions where no direct training data is available for the task at hand. Instead of learning from examples, zero-shot learning models use pre-existing knowledge and semantic understanding to classify or identify objects they have never explicitly been trained on. This method is particularly powerful in scenarios where it's impractical to have training data for every possible category or outcome, allowing models to infer and make predictions about entirely unseen data.

Each of these approaches addresses the challenge of data scarcity from a different angle, offering solutions that range from utilizing a handful of examples to making educated guesses without any examples at all. Through these methodologies, machine learning can achieve remarkable flexibility and adaptability, pushing the boundaries of what's possible even when data is limited.

Examples

Sure, here are the examples of few-shot learning, one-shot learning, and zero-shot learning in the context of large language models.

Few-Shot Learning

- **Scenario:** Providing a language model with a small number of examples to perform a specific task.

- **Example:** You want a language model to generate poetry in the style of a specific poet. You provide it with a few poems (e.g., three to five poems) written by that poet.

- **Training Data:** Poet A – three to five poems

- **Task:** After seeing these few examples, the language model can generate new poems that mimic the style and themes of Poet A.

One-Shot Learning

- **Scenario:** Providing a language model with a single example to learn a task.

- **Example:** You want a language model to translate a specific phrase from English to a rare language it has never encountered. You provide it with one example translation.

- **Training Data**
  - Phrase in English: "Good morning"
  - Phrase in Rare Language: "Buenos días"

- **Task:** After seeing this single example, the language model can translate other similar phrases from English to the rare language.

Zero-Shot Learning

- **Scenario:** Asking a language model to perform a task it hasn't been explicitly trained on by leveraging its general knowledge.

- **Example:** You ask a language model to summarize a scientific article on a topic it has never seen before, using its understanding of language and general knowledge about science.

- **Training Data:** General knowledge of scientific concepts and language patterns.

- **Auxiliary Information for Zero-Shot Learning:** The model uses its pre-existing knowledge about scientific articles and summarization techniques.

- **Task:** Summarize a new scientific article accurately, even though the model hasn't seen this specific article or topic before.

These examples illustrate how large language models can be adapted to new tasks and contexts with varying amounts of specific training data or even just descriptive information, showcasing the flexibility and power of modern AI systems.

LLM Hallucinations

AI hallucination refers to the occurrence where advanced artificial intelligence systems, particularly large language models (LLMs) and computer vision technologies, generate outputs that contain fabricated or nonsensical information, not grounded in reality or the data they were trained on. This phenomenon can result in responses or visual outputs that seem bizarre or entirely incorrect from a human perspective.

Typically, users expect AI-generated responses to accurately reflect the information pertinent to their queries or prompts. Nonetheless, there are instances where the AI's algorithms produce outcomes that diverge from the underlying training data, misinterpret the input due to flawed decoding processes, or generate responses without a discernible logical foundation, effectively "hallucinating" the information.

The use of the term "hallucination" to describe such AI behaviors might appear odd at first, as it anthropomorphizes machine processes with a term usually applied to human or animal experiences. Yet this metaphor aptly conveys the unexpected and often surreal nature of the outputs, reminiscent of how humans might discern shapes in clouds or faces on the moon, driven by the AI's misinterpretations caused by issues like overfitting, biases or inaccuracies in the training data, and the inherent complexity of the models themselves.

Instances of AI hallucinations have been noted in several high-profile cases, underscoring the challenges inherent in deploying generative, open source AI technologies. Examples include Google's Bard chatbot making unfounded claims about astronomical discoveries, Microsoft's AI expressing inappropriate emotional attachments or behaviors, and Meta withdrawing its Galactica LLM demo due to the propagation of biased or incorrect information.

Although steps have been taken to address and rectify these problems, they highlight the potential for unexpected and sometimes problematic outcomes in the application of AI technologies, even under optimal conditions.

Classification of Hallucinations in Large Language Models (LLMs)

LLMs exhibit two primary types of hallucinations, differentiated by their nature and impact: factuality hallucinations and faithfulness hallucinations.

Factuality Hallucinations

These occur when an LLM produces information that is factually incorrect. An example of this would be an LLM asserting that Charles Lindbergh was the first person to land on the moon, clearly a factual mistake. Such errors typically stem from the model's inadequate grasp of context and the presence of inaccuracies or misleading information in its training data, resulting in outputs that fail to align with actual facts.

Faithfulness Hallucinations

This category encompasses situations where the content generated by an LLM deviates from or contradicts the source material it is supposed to reflect or summarize. For instance, during summarization tasks, if a source article mentions the FDA's approval of the first Ebola vaccine in 2019, a faithfulness hallucination might manifest as the model inaccurately stating that the FDA disapproved of the vaccine (an intrinsic hallucination), or it might introduce unrelated information, such as the development of a COVID-19 vaccine by China (an extrinsic hallucination), neither of which are supported by the source text.

These classifications highlight the challenges LLMs face in accurately interpreting and reproducing information, underscoring the importance of ongoing efforts to enhance their understanding and reliability.

Implications of AI Hallucination

The repercussions of AI hallucination are profound, especially in critical sectors such as healthcare, where an AI misdiagnosis could mistakenly identify a harmless condition as severe, prompting unnecessary treatments. Moreover, AI-generated inaccuracies can fuel the dissemination of false information. Consider a scenario where AI-driven news platforms inaccurately report on an unfolding crisis based on unchecked facts, potentially exacerbating the situation by spreading misinformation.

A primary factor contributing to AI hallucinations is the presence of bias in the training data. When AI systems are trained on datasets that are skewed or not fully representative, they might generate outputs that inaccurately reflect these biases, interpreting nonexistent patterns or characteristics.

Furthermore, AI systems are susceptible to adversarial attacks, where malicious entities deliberately alter inputs to deceive the AI into making incorrect identifications or decisions. In the context of image recognition, such an attack could involve the introduction of imperceptible, specially designed noise to an image, leading the AI to erroneously categorize it. This vulnerability is particularly alarming in areas critical to public safety and security, including cybersecurity measures and the development of autonomous driving technologies.

To counter these threats, AI researchers are diligently working on developing robust defense mechanisms, such as adversarial training, which involves training AI models on both standard and adversarially modified inputs to enhance their resilience against such attacks. Despite these advancements, maintaining rigorous standards during the training process and ensuring thorough verification of information remain essential to mitigating the risks associated with AI hallucinations.

There are several reasons why LLM hallucinations occur:

- **Training Data Limitations:** LLMs are trained on vast datasets compiled from the Internet and other sources, which might contain inaccuracies, biases, or outdated information. The model may replicate these inaccuracies in its responses.

- **Inference and Generalization:** LLMs generate responses based on patterns and associations learned during training. In attempting to provide coherent and contextually appropriate answers, they might infer details or make generalizations that are incorrect.

- **Lack of Real-World Understanding:** While LLMs can process and generate language at a level that often seems understanding, they don't possess true comprehension or awareness. They operate based on statistical relationships between words, which can lead to plausible but entirely fabricated statements.

- **Complexity of Language and Knowledge:** Language is inherently ambiguous and complex. An LLM might struggle with nuanced or complex topics, leading to simplifications or errors that manifest as hallucinations.

Efforts to mitigate LLM hallucinations include improving the models' training data, refining their architectures, and developing more sophisticated techniques for checking and validating generated content. Additionally, user feedback and prompt engineering (designing inputs to the model that are more likely to yield accurate and relevant outputs) are crucial strategies for reducing the occurrence of hallucinations.

Mitigating the Risks of AI Hallucinations: Strategies for Prevention

To effectively curb the occurrence of AI hallucinations, proactive measures must be implemented to ensure AI models operate within the bounds of accuracy and reliability. Here are essential strategies to maintain the integrity of AI outputs.

Ensure High-Quality Training Data

The foundation of generative AI's performance lies in the training data. To avert hallucinations, it's crucial to utilize training datasets that are comprehensive, diverse, and meticulously curated. High-quality data aids in reducing biases and enhancing the model's understanding of its tasks, leading to more accurate and reliable outputs.

Clarify the Model's Purpose and Constraints

Defining the specific objectives and limitations of your AI model is critical in minimizing irrelevant or incorrect outputs. Establishing clear guidelines on the model's intended use helps focus its capabilities on generating pertinent results, thereby reducing the chances of hallucinations.

Implement Data Templates

Utilizing data templates standardizes the format for input data, guiding the AI model toward producing outputs that adhere to established norms. This approach promotes consistency in the model's responses and helps prevent deviations into inaccurate or irrelevant territory.

Restrict Possible Outcomes

Constraining the range of an AI model's responses can significantly reduce hallucinations. By setting explicit boundaries through filtering mechanisms or probability thresholds, you can enhance the accuracy and consistency of the model's outputs.

Continuous Testing and Refinement

Rigorous testing prior to deployment and ongoing evaluation of the AI model are essential for maintaining its effectiveness. Regular assessments allow for timely adjustments or retraining as needed, ensuring the model remains current and accurate as new data emerges.

Incorporate Human Oversight

Integrating human review into the AI output validation process serves as an essential safeguard against hallucinations. Human oversight brings an additional layer of scrutiny, with the capacity to identify and rectify inaccuracies that the AI might overlook. Expert reviewers can also contribute valuable insights, further enhancing the model's accuracy and relevance.

When Hallucinations Might Be Good?

LLM (large language model) hallucinations refer to instances where the model generates information that is not factually accurate or completely fabricated. While generally seen as a downside, there are contexts where these hallucinations can be beneficial:

- **Creative Writing:** Hallucinations can spur creativity and generate unique ideas, storylines, or characters that a human writer might not conceive. This can be especially useful in brainstorming sessions for fiction writing, screenwriting, or game design.

- **Artistic Expression:** In poetry, music lyrics, and other forms of artistic expression, the generation of unexpected and surreal imagery can enhance the creative process and lead to innovative art forms.

- **Brainstorming and Ideation:** During the initial stages of brainstorming, hallucinations can provide a wide range of ideas, including out-of-the-box concepts that might not emerge from a purely factual approach.

- **Role-Playing and Simulation:** In gaming and virtual environments, hallucinations can create dynamic and unpredictable scenarios, enhancing the immersive experience for players.

- **Educational Purposes:** In certain educational contexts, hallucinations can stimulate critical thinking and analysis. For instance, students might be tasked with identifying inaccuracies or fabrications in a generated text, thereby honing their research and fact-checking skills.

- **Therapeutic Uses:** Creative writing and storytelling are often used in therapeutic settings. Hallucinated content can help patients explore their thoughts and feelings in novel ways, providing new perspectives and insights.

- **Exploration of Alternate Realities:** For science fiction and speculative fiction, hallucinations can help in constructing alternative realities, future scenarios, or parallel universes, enriching the narrative with imaginative possibilities.

While hallucinations have these potential benefits, it is crucial to manage them carefully to avoid misinformation in contexts where accuracy is paramount, such as news, medical advice, and legal information.

Future Implications

Although large language models (LLMs) hold the promise of transforming numerous sectors, it's crucial to acknowledge their constraints and the ethical concerns they raise. Companies and professionals ought to weigh the benefits against the potential drawbacks and hazards of implementing LLMs. Moreover, it's imperative for creators of these models to persistently enhance their designs, aiming to reduce biases and augment their applicability across varied contexts.

As we navigate through the existing boundaries of large language models (LLMs), the quest for the next breakthrough in AI is leading researchers down innovative paths. A key insight fueling this journey is the realization that the majority of data interacted with daily by humans is scarcely digital, and an even smaller fraction is textual.

The exploration into multimodal learning stands out as a pivotal direction, merging textual data with other forms such as images, videos, and audio. This integration promises to unlock a richer, more complex understanding of information, enabling AI systems to grasp and interpret human language with unprecedented depth and nuance.

Achieving this enhanced understanding necessitates progress in specialized fields including computer vision and video analysis, potentially revolutionizing speech recognition for more engaging and holistic AI interactions. However, the shift toward a multimodal approach introduces its own set of challenges, particularly in handling the increased data scale and complexity, necessitating innovative solutions to streamline and optimize the training process.

In response to these challenges, the development of more efficient, environmentally friendly training methodologies has become a focal point. Techniques such as few-shot learning, transfer learning, and meta-learning are being harnessed to diminish the dependency on vast datasets and computing power, facilitating a more sustainable and accessible AI landscape. These strategies leverage existing knowledge, applying it across various domains to foster AI applications that are not only high performing but also mindful of energy use and environmental impact.

Parallel to these efforts, there's a growing emphasis on enhancing AI's contextual and generalization capabilities through the integration of Symbolic AI with deep learning, giving rise to Hybrid AI. This approach seeks to marry the data-driven insights of connectionism with the structured reasoning of Symbolic AI, a paradigm once sidelined due to its scalability challenges but now recognized for its potential to synergize with statistical models.

Hybrid AI aims to automate the extraction of symbolic rules from vast datasets, transforming the daunting task of manually defining these systems into a streamlined, machine-led process. This fusion not only elevates AI's cognitive and problem-solving capacities but also addresses the pressing issues of explainability and interpretability inherent in deep learning models by grounding decisions in clearly defined rules and logic.

Looking to the future, while the current enthusiasm in AI circles is largely centered around generative models and LLMs, signs suggest that their performance may soon reach a saturation point. The horizon of AI, however, stretches far beyond, with multimodal learning, sustainable practices, and Hybrid AI poised to define the next generation of AI systems. These advancements promise to usher in an era of AI that is more versatile, sustainable, and capable of tackling a broader spectrum of challenges with greater efficacy and generalizability.

Examples of LLM Architectures

Large language models (LLMs) have revolutionized natural language processing with their ability to understand and generate human-like text. Key examples include GPT (Generative Pre-trained Transformer), developed by OpenAI, which uses transformer architecture to generate coherent and contextually relevant text based on given prompts. Another notable example is BERT (Bidirectional Encoder Representations from Transformers), created by Google, which focuses on understanding context in both directions (left and right) for tasks like question answering and sentiment analysis. Additionally, T5 (Text-to-Text Transfer Transformer), also by Google, converts all NLP tasks into a text-to-text format, enabling a unified approach to various language tasks. These architectures leverage the power of transformers to achieve state-of-the-art performance in diverse NLP applications.

GPT-4

The fourth iteration in the series of foundational models by OpenAI, the Generative Pre-trained Transformer 4 (GPT-4), is a multimodal language model that was introduced on March 14, 2023. It is accessible through the subscription-based ChatGPT Plus, OpenAI's API, and Microsoft Copilot, a free chatbot service. GPT-4 employs a transformer architecture, leveraging a combination of publicly available data and data obtained under license from third-party providers for pre-training. This process involves predicting subsequent tokens, which is then refined through fine-tuning with human and AI-generated reinforcement learning feedback to ensure alignment with human values and adherence to policy guidelines.

Compared to its predecessor, GPT-3.5, the GPT-4 version of ChatGPT is seen as an enhancement, though it still shares some limitations of its earlier versions. A distinctive feature of GPT-4, referred to as GPT-4V, includes the **ability to process image inputs in ChatGPT**. Despite its advancements, OpenAI has chosen not to disclose specific technical details and metrics about the model, including its exact size.

Model Characteristics

- **Scale and Architecture of GPT-4:** GPT-4 boasts approximately **1.8 trillion parameters** distributed over **120 layers**, significantly surpassing GPT-3 by more than a factor of ten.

- **Expertise Integration Through MoE:** The architecture employs 16 specialized experts, each with around **111 billion parameters** dedicated to multilayer perceptrons (MLP). For each forward pass, two experts are selected, optimizing cost efficiency.

- **Training Dataset Composition:** The model's training regimen encompassed around 13 trillion tokens, sourced from a mix of textual and coding data, supplemented by fine-tuning contributions from **ScaleAI** and internal resources.

- **Diversity of Training Data:** Its dataset amalgamated content from **Common Crawl and RefinedWeb**, among other speculated sources such as **Twitter, Reddit, YouTube**, and extensive textbook collections, amounting to a total of 13 trillion tokens.

- **Cost of Development:** The financial investment in GPT-4's development reached approximately $63 million, reflecting the extensive computational resources and time commitment required.

- **Operational Expenses for Inference:** Operating GPT-4 incurs thrice the expense of its predecessor, the 175-billion-parameter Davinci model, attributed to the necessity for larger computing clusters and reduced efficiency in resource utilization.

- **Inference Process:** The model's inference mechanism operates over a network of 128 GPUs, leveraging 8-way tensor parallelism and 16-way pipeline parallelism for efficient processing.

- **Enhanced Vision Capabilities:** GPT-4 integrates a vision encoder, akin to the Flamingo architecture, enabling the model to interpret web pages and transcribe imagery and video content. This feature introduces additional parameters and is further refined with about 2 trillion tokens of fine-tuning data, enriching its multimodal capabilities.

- **Training Compute:** Trained on ~25,000 Nvidia A100 GPUs over 90–100 days.

- **Training Data:** Trained on a dataset of ~13 trillion tokens.

- **Context Length:** Supports up to 32,000 tokens of context.

GPT-4 is capable of handling over **25,000 words of text**, allowing for use cases like long form content creation, extended conversations, and document search and analysis.

GPT-4 Limitations

Similar to its forerunners, GPT-4 sometimes produces information that is either not present in its training data or contradicts the input provided by users, a phenomenon often referred to as "hallucination." Additionally, the model operates with a lack of transparency regarding its decision-making process. Although it can offer explanations for its responses, these are generated after the fact and may not accurately represent the underlying reasoning process. Frequently, the explanations provided by GPT-4 can be inconsistent with its prior responses.

In an evaluation using **ConceptARC**,[3] a benchmark designed for assessing abstract reasoning, GPT-4's performance was significantly lower than expected, scoring under 33% across all categories. This was in stark contrast to specialized models and human performance, which scored around 60% and at least 91%, respectively.

This outcome suggests that abstract reasoning, which involves understanding complex relationships and patterns, remains a challenging domain for general-purpose AI models like GPT-4. This benchmark focuses on abstract reasoning, a critical aspect of human cognition that involves identifying patterns, logical rules, and relationships among objects. GPT-4's sub-33% performance indicates its struggles in this domain.

The specialized models, designed with specific architectures or trained on datasets tailored for abstract reasoning, showed much better performance, scoring at 60%, emphasizing the importance of domain-specific training and optimization.

Specialized models refer to those specifically designed and optimized to handle particular types of tasks or domains. In the context of abstract reasoning and the ConceptARC benchmark, specialized models demonstrated superior performance compared to GPT-4. Here are details about these specialized models.

Table 5-1 compares the performance of humans, the top two ARC-Kaggle competition entries, and GPT-4 on various conceptual tasks. Each task is evaluated based on the accuracy or success rate of completing the task.

---

[3] Arseny Moskvichev, Victor Vikram Odouard, Melanie Mitchell, The ConceptARC Benchmark: Evaluating Understanding and Generalization in the ARC Domain, https://arxiv.org/abs/2305.07141

CHAPTER 5   BASIC OVERVIEW OF THE COMPONENTS OF THE LLM ARCHITECTURES

Table 5-1. *Performance comparison across conceptual tasks*[4]

Concept	Humans	ARC-Kaggle First Place	ARC-Kaggle Second Place	GPT-4
Above and Below	0.90	0.70	0.33	0.23
Center	0.94	0.50	0.20	0.33
Clean Up	0.97	0.50	0.20	0.20
Complete Shape	0.85	0.47	0.30	0.23
Copy	0.94	0.23	0.27	0.23
Count	0.88	0.60	0.40	0.13
Extend To Boundary	0.93	0.77	0.47	0.07
Extract Objects	0.86	0.43	0.43	0.03
Filled and Not Filled	0.96	0.73	0.43	0.17
Horizontal and Vertical	0.91	0.43	0.10	0.27
Inside and Outside	0.91	0.57	0.10	0.10
Move To Boundary	0.91	0.37	0.30	0.20
Order	0.83	0.27	0.23	0.27
Same and Different	0.88	0.53	0.17	0.17
Top and Bottom 2D	0.95	0.60	0.57	0.23
Top and Bottom 3D	0.93	0.50	0.03	0.20

1. **ARC-Kaggle Challenge Winners**: The first- and second-place programs from the ARC-Kaggle challenge were evaluated on the ConceptARC benchmark. These models were designed explicitly for solving the Abstraction and Reasoning Corpus (ARC) tasks, which involve abstract reasoning similar to those in ConceptARC. Their design likely includes algorithms optimized for recognizing and manipulating abstract patterns and concepts.

2. **Human Performance**: For context, human participants, tested on the same tasks via platforms like Amazon Mechanical Turk and Prolific, consistently outperformed both GPT-4 and the specialized models, highlighting the complexity of abstract reasoning tasks and the gap between current AI capabilities and human cognition.

---

[4] Arseny Moskvichev, Victor Vikram Odouard, Melanie Mitchell, The ConceptARC Benchmark: Evaluating Understanding and Generalization in the ARC Domain, https://arxiv.org/abs/2305.07141

The specialized models (first and second place in ARC-Kaggle) outperformed GPT-4 across various concept groups in the ConceptARC benchmark. Here are some examples of their performance.

- **Above and Below**: First Place ARC-Kaggle scored 70%, while GPT-4 scored 23%.

- **Extend to Boundary**: First Place ARC-Kaggle scored 77%, compared to GPT-4's 7%.

- **Filled and Not Filled**: First Place ARC-Kaggle scored 73%, while GPT-4 scored 17%.

Key Takeaways

- **Optimization for Specific Tasks**: The success of the specialized models highlights the importance of tailoring algorithms to the specific requirements of abstract reasoning tasks, including the ability to generalize from specific instances of core concepts.

- **Limitations of General-Purpose Models**: GPT-4, while powerful, is a general-purpose model not specifically optimized for tasks like those in ConceptARC. This illustrates the need for continued development in AI to bridge the gap in abstract reasoning capabilities.

Sam Bowman, a researcher not involved in the study, noted that these findings might not necessarily point to a deficiency in abstract reasoning capabilities of GPT-4, considering that the assessment was visually oriented and GPT-4 is primarily a linguistic model.

A study in January 2024 by Cohen Children's Medical Center[5] researchers reported that GPT-4 had a 17% accuracy rate in diagnosing pediatric medical conditions, highlighting limitations in its medical diagnostic capabilities.

Regarding bias, GPT-4 underwent a two-stage training process. Initially, it was fed vast amounts of Internet text to learn to predict the next token in a sequence. Subsequently, it underwent refinement through reinforcement learning from human

---

[5] *JAMA Pediatr.* Published online, January 2, 2024. doi: 10.1001/jamapediatrics.2023.5750

feedback, aimed at teaching the model to reject prompts that could lead to harmful behaviors, as defined by OpenAI. This includes generating responses related to illegal activities, self-harm, or the description of explicit content.

Researchers from Microsoft have raised concerns that GPT-4 may display certain cognitive biases, including confirmation bias, anchoring, and base-rate neglect, indicating potential areas where the model's reasoning could be influenced by inherent biases.

BERT

BERT, short for Bidirectional Encoder Representations from Transformers, is a freely available software library developed for tasks related to natural language processing (NLP). Introduced in 2018 by the **Google AI Language team**, BERT is designed to understand the subtleties and context of human language.

Introduction to BERT

At its core, BERT utilizes a neural network architecture based on transformers to process and interpret the nuances of human language. Unlike the traditional transformer model, which includes both encoder and decoder components, BERT incorporates only the encoder mechanism. This design choice underscores **BERT's focus on comprehending input texts over generating new text**.

The Bidirectional Nature of BERT

Conventional language models typically analyze text in a linear fashion, processing words one after the other in either forward or backward direction. This approach can overlook the full context surrounding a word. Contrary to this, BERT adopts a **bidirectional strategy**, evaluating the context on both sides of a word within a sentence simultaneously. This method enables BERT to gain a comprehensive understanding of the text by considering the entire sentence context at once.

Take the sentence: "She went to the bank to _____ some money." In models that process text from one direction, the meaning of the gap relies heavily on the words leading up to it, possibly confusing whether "bank" signifies a riverbank or a financial institution. BERT's bidirectional feature, however, evaluates the context preceding ("She went to the bank to") and following ("some money") the blank space in tandem. This approach

CHAPTER 5   BASIC OVERVIEW OF THE COMPONENTS OF THE LLM ARCHITECTURES

furnishes a more layered interpretation, recognizing that the "bank" here pertains to a financial establishment, given the context of money. This showcases the enhanced comprehension BERT achieves by considering the entire sentence's context, overcoming limitations seen in unidirectional models.

Training Stages of BERT: Pre-training and Fine-Tuning

BERT's development involves a dual-phase strategy:

- **Pre-training on vast volumes of unlabeled text data** to master contextual relationships between words
- **Fine-tuning with task-specific labeled data** to apply this understanding to particular NLP challenges

Phase 1: Pre-training with Unlabeled Data

During its pre-training, BERT is exposed to extensive corpora of text without specific labels. It learns rich word embeddings that **reflect the words' meanings based on their contexts**. This phase includes self-directed tasks, such as filling in blanks within sentences (a technique known as Masked Language Modeling, or MLM) and determining whether two sentences logically follow one another. These exercises enable BERT to capture the subtle nuances of language and sentence structure.

Phase 2: Fine-Tuning for Specific Tasks

Once pre-trained, BERT is fine-tuned by adjusting its parameters on a smaller, task-specific dataset. This process refines BERT's broad language understanding to perform specific tasks like sentiment analysis, question answering, or entity recognition. The model's versatile architecture allows for easy adaptation across various NLP applications with only minor adjustments needed for each new task.

How BERT Functions

BERT leverages just the encoder part of the transformer architecture, processing input token sequences to generate contextualized token embeddings (Figure 5-13). Unlike traditional models that predict the next word linearly, BERT employs two novel pre-training objectives:

241

CHAPTER 5 BASIC OVERVIEW OF THE COMPONENTS OF THE LLM ARCHITECTURES

- **Masked Language Model (MLM):** BERT hides a fraction of the input tokens at random and then predicts these masked tokens based on their context, enhancing its grasp of language contextuality.

- **Next Sentence Prediction (NSP):** BERT assesses whether two sentences are logically sequenced, further refining its understanding of narrative flow.

BERT's Architectural Innovations

The architecture of BERT, whether in its BASE or LARGE variant, expands upon the original transformer model (Figure 5-14) with more encoder layers, larger feedforward networks, and more attention heads, enabling it to capture a deeper level of linguistic detail. The BASE model incorporates 12 transformer blocks (with 768 hidden units and 12 attention heads), while the LARGE model doubles this capacity.

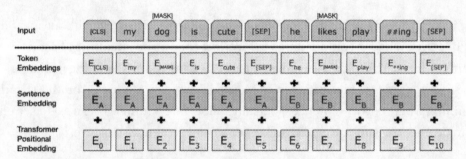

Figure 5-13. BERT input embeddings visualization (Source: http://medium.com)

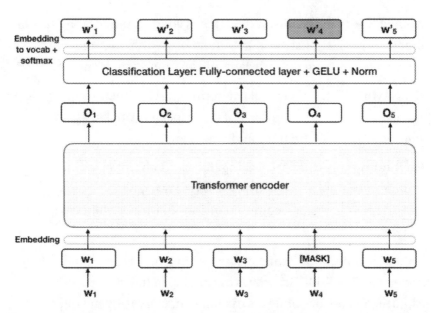

Figure 5-14. *Transformer encoder classification layer (Source:* http://medium.com*)*

From Training to Application

BERT processes inputs by starting with a special classification token (CLS) followed by the input text. The model's layers apply self-attention and feedforward networks sequentially, producing a vector representation for each token. The CLS token's final representation can serve as a feature for classification tasks, illustrating BERT's ability to transform its pre-trained embeddings into task-specific applications, such as single-layer neural network classifications, with remarkable success.

Uses of BERT in Language Processing

BERT's capabilities are harnessed across a variety of language understanding and generation tasks, such as the following:

- **Generating Word Embeddings:** BERT excels in creating detailed representations of words within their sentence context, enhancing the understanding of their meanings and relationships.

- **Named Entity Recognition (NER):** It is adept at pinpointing specific names, organizations, locations, and more within text, making it invaluable for extracting relevant information from documents.

- **Classifying Text:** For tasks like determining sentiment, identifying spam, or categorizing topics, BERT's deep contextual insights significantly boost accuracy and efficiency.

- **Facilitating Question Answering:** By comprehending questions and sourcing accurate answers from provided texts, BERT enhances the effectiveness of reading comprehension aids and automated inquiry systems.

- **Enhancing Machine Translation:** Leveraging its contextual understanding, BERT improves the quality of translations by capturing linguistic subtleties essential for conveying meaning accurately across languages.

- **Summarizing Texts:** In abstracting key points from extensive texts to produce coherent and relevant summaries, BERT's comprehension of context and semantics shines.

- **Powering Conversational Interfaces**: Whether in chatbots, digital assistants, or interactive systems, BERT's nuanced language interpretation enables more fluid and natural dialogues.

- **Evaluating Semantic Similarity:** By comparing the BERT-generated embeddings of texts, it's possible to evaluate their semantic closeness, aiding in identifying paraphrases, detecting duplicates, and retrieving information based on meaning similarity.

Through these applications, BERT significantly advances the field of natural language processing, offering sophisticated solutions to previously challenging problems.

T5

The "Text-to-Text Transfer Transformer" (T5) (Figure 5-15), unveiled by Google in 2020,[6] represents an advanced model framework based on the transformer architecture, specifically employing both encoder and decoder components for generating text. This approach distinguishes it from other notable transformer-based models like BERT or GPT, which utilize either encoder or decoder structures but not both. The innovation of the T5 model is further underscored by its introduction of the Colossal Clean Crawled Corpus (C4), a vast, meticulously cleaned dataset designed for pre-training the language model through self-supervised learning techniques.

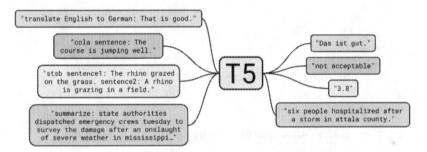

Figure 5-15. T5 model tasks and outputs (Source: Google Blog)

Training or fine-tuning the T5 model necessitates a pair of input and output text sequences, enabling the execution of diverse tasks such as text generation, translation, and more. Subsequent developments have led to various iterations of the original T5 model, including T5v1.1, which benefits from architectural improvements and exclusive pre-training on the C4 dataset; mT5, a multilingual variant trained across 101 languages; ByT5, which leverages byte sequences instead of subword token sequences; and LongT5, tailored for processing extended text inputs.

The T5 model's examination of transformer architectures (Figure 5-16) revealed three main types: the standard encoder-decoder, the single-layer language model, and the prefix language model, each distinguished by unique masking strategies to control attention mechanisms. Among these, the encoder-decoder setup, characterized by its comprehensive masking techniques, proved to be the most effective.

---

[6] Colin Raffel, Noam Shazeer, Adam Roberts, Katherine Lee, Sharan Narang, Michael Matena, Yanqi Zhou, Wei Li, Peter J. Liu, Exploring the Limits of Transfer Learning with a Unified Text-to-Text Transformer

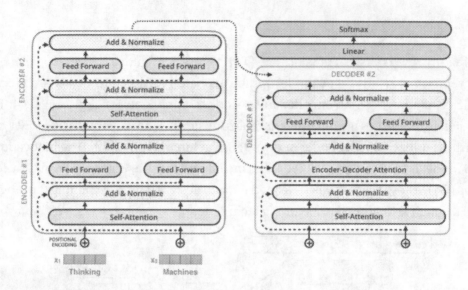

Figure 5-16. Transformer encoder-decoder architecture (Source: Jay Alamar's Blog)

T5's pre-training strategy incorporates a multitask learning approach, dividing tasks into unsupervised and supervised learning categories. Unsupervised tasks involve training on the C4 dataset with a focus on maximizing likelihood objectives, while supervised tasks encompass a range of NLP applications, all reformatted to fit the text-in, text-out paradigm essential for encoder-decoder models. This methodological diversity, alongside the introduction of special tokens for formatting text sequences, facilitates versatile model training.

Explorations into unsupervised training objectives highlighted the effectiveness of different strategies, with a BERT-style masking approach emerging as superior. This entails masking specific words or spans within the text for the model to predict, a method refined through experimentation with various corruption rates and span lengths to optimize performance.

Cohere

Cohere has emerged as a distinguished entity within the field of natural language processing (NLP), achieving remarkable progress since its foundation. The organization is driven by a vision to master language comprehension, which has culminated in the creation of an advanced language model. A pivotal achievement for Cohere has been the integration of transformer technologies, enhancing the model's capability to analyze and generate text with an acute awareness of context and meaning. Additionally, Cohere places a strong emphasis on the ethical deployment of AI, instituting comprehensive measures to counteract bias and adhere to ethical principles, thereby ensuring the development of both potent and principled AI tools.

At the heart of Cohere's technological infrastructure lies the implementation of cutting-edge neural network innovations, particularly transformer models. These models excel in processing sequential information, grasping the intricate dynamics between words within and across sentences. Employing attention mechanisms, Cohere's models adeptly prioritize significant segments of text when necessary. This meticulous attention to context and sequence grants Cohere a superior capacity for deciphering the subtleties of language, including its tone, style, and underlying implications. Designed with scalability in mind, the architecture effortlessly accommodates an array of linguistic tasks, ranging from straightforward text categorization to elaborate question-answering frameworks.

Known for their effectiveness and scalability, Cohere's models deliver consistent results even under limited computational conditions. These models are exceptionally advantageous for enterprises seeking a tailor-made solution that seamlessly meshes with their current infrastructure without demanding extensive computational power. Cohere's proficiency in language comprehension and generation renders it an excellent tool for automating customer support, analyzing sentiments, and crafting content.

In various sectors, Cohere's AI model demonstrates its versatility. In customer support, for instance, it powers chatbots capable of not just understanding and addressing user inquiries but also tailoring responses to reflect the customer's emotional state and prior interactions. When it comes to content moderation, Cohere plays a crucial role in efficiently screening and managing user-generated content to ensure adherence to community standards. Moreover, in the educational technology space, Cohere's models are instrumental in customizing learning materials to fit individual preferences and learning velocities, thereby transforming the e-learning landscape.

PaLM 2

Language models have revolutionized the field of natural language processing, significantly enhancing AI's capacity to understand and produce text that closely mimics human communication. Among these innovative developments, the Pathways Language Model 2 (PaLM 2) is a standout example, advancing the frontiers of linguistic comprehension and context-aware processing.

This guide offers an in-depth exploration of PaLM 2, examining its structure, functions, and the groundbreaking approaches it employs to achieve unparalleled mastery over language. Building on the groundwork of its forerunner, PaLM, this iteration introduces cutting-edge methodologies that have transformed the landscape of natural language understanding.

Embark on a journey with us as we illuminate the complexities of PaLM 2, revealing the future directions of language modeling.

How PaLM 2 Operates

Grasping the functionality of PaLM 2 involves a deep dive into its foundational technology and components. Here's a breakdown of how PaLM 2 operates.

Initial Data Acquisition and Preparation

PaLM 2 begins with collecting a vast, diverse dataset from a variety of sources, including books, articles, websites, and social media, to cover a broad spectrum of language use. This data is then meticulously cleaned and preprocessed to remove any irrelevant content and noise, followed by tokenization to break the text into manageable units and sentences. This step ensures the data is uniform and primed for analysis.

Leveraging Transformer Architecture

At its core, PaLM 2 utilizes the groundbreaking transformer architecture, which has revolutionized NLP by introducing self-attention mechanisms. These mechanisms enable the model to assess the significance of words within a context, enhancing its ability to make precise predictions and deepen its understanding of the language. The architecture's efficiency in training and parallel processing capabilities make it ideal for managing extensive language models like PaLM 2.

Extensive Pre-training

The pre-training phase involves the model learning through predicting missing words, grasping contexts, and generating coherent text across a vast dataset. This exposure allows PaLM 2 to familiarize itself with various language patterns and nuances, progressively honing its linguistic representation skills.

Task-Specific Fine-Tuning

While general pre-training provides a broad language foundation, fine-tuning tailors PaLM 2 for specific applications by training it on targeted, domain-specific datasets. This process enables the model to apply its extensive language understanding to specific real-world tasks effectively.

The Novel Pathways Architecture

What sets PaLM 2 apart is its unique Pathways architecture, featuring multiple independent pathways for processing different linguistic elements. This design allows for specialized attention to various aspects of language, enhancing the model's overall comprehension abilities.

Independent Pathway Functioning

Each pathway operates autonomously, focusing on specific linguistic features without cross-interference. This independence allows for dedicated processing, such as syntactic analysis or semantic understanding, leading to a richer interpretation of text.

Adaptive Computational Allocation

PaLM 2 dynamically adjusts its computational resource allocation based on the complexity of the input, ensuring efficient and accurate processing for both simple and complex queries.

Pathway Interaction and Collaboration

Despite their independence, the pathways in PaLM 2 are designed to interact and share information, fostering a comprehensive understanding of language by leveraging the strengths of each pathway.

Selective Pathway Engagement

Upon receiving input, PaLM 2 evaluates and selects the most appropriate pathway for processing, optimizing its response accuracy and relevance to the specific linguistic challenge at hand.

Generating Outputs

After processing, PaLM 2 produces outputs tailored to the task it's fine-tuned for, showcasing its versatility across a range of language processing applications.

PaLM 2 signifies a monumental stride in AI, ushering in a new era of sophisticated language understanding and generation. By incorporating advanced techniques and a multifaceted architecture, PaLM 2 excels in adaptability and generalization, establishing itself as a formidable tool for addressing complex linguistic tasks.

With its profound grasp of context and nuanced expression, PaLM 2 promises more natural and human-like interactions with AI systems, enhancing user experiences across various applications. As we move forward, the impact of PaLM 2 on the development of conversational agents, machine translation, and text summarization will undoubtedly be profound, marking a significant milestone in the evolution of AI technologies.

Jurassic-2

AI21's Jurassic-2 language model comes in three variants – Jumbo, Grande, and Large – with each offered at distinct price levels. The specifics of the model sizes are kept confidential; however, the documentation highlights the Jumbo version as the most potent option. These models are characterized as versatile, excelling across all types of generative tasks. The Jurassic-2 model is proficient in seven languages and allows for fine-tuning with specific datasets. Users can obtain an API key through the AI21 platform and utilize the AI21() class for model access.

The J2 models have been developed using an extensive database of textual content, equipping them with the capability to generate text that closely mimics human writing. They excel in a wide range of complex activities, including but not limited to answering questions, categorizing text, and more.

These models are adaptable to almost any task involving language, through the use of prompt engineering. This involves designing a prompt that outlines the task at hand and may include examples. They are particularly beneficial for creating advertising content, powering conversational agents, and enhancing creative writing efforts.

While experimenting with different prompts can lead to satisfactory outcomes for your specific needs, optimizing performance and expanding your application's capabilities may require training a bespoke model.

Claude v1

Claude v1, Anthropic's inaugural release of its conversational AI assistant, marks a significant achievement in the company's quest to create safe artificial general intelligence. Established by Dario Amodei and Daniela Amodei in 2021, after their tenure at OpenAI focusing on AI safety, Anthropic is a San Francisco-based AI safety startup. The organization is committed to crafting AI technologies that are beneficial, nonharmful, and truthful, emphasizing the importance of aligning AI development with human values through a safety and ethics-first research approach.

In April 2022, Anthropic introduced Claude v1 to the public, following more than a year of development in stealth. This debut was designed to demonstrate the company's commitment to delivering safe and effective conversational AI tools. Claude is engineered to engage in natural conversations while steering clear of damaging, unethical, or false exchanges.

Data and Training Approach

The foundation of Claude v1's training involved a method known as Constitutional AI, which leverages a vast and varied dataset of natural dialogues that embody a range of personas and communicative styles, intentionally omitting any harmful content. Anthropic developed a unique training methodology named Vigilance to embed safety mechanisms into Claude, encouraging the system to learn and adhere to social norms and ethical guidelines through social learning and evaluative feedback.

Claude's initial training utilized supervised learning methods to refine the model's ability to generate beneficial and nondetrimental responses. This was complemented by self-supervised learning to enhance Claude's broader conversational aptitudes.

Model Design

Anthropic crafted a bespoke neural network structure for Claude v1, based on transformer technology commonly applied in natural language processing tasks. While the precise size of Claude v1's model is proprietary, its billions of parameters enable it to handle extensive discussions across myriad subjects.

Incorporating safety measures such as Constitutional AI and Vigilance into its architecture, Claude v1 is engineered to avoid unsafe interactions. Its primary features include engaging in natural dialogues, providing valuable assistance, refraining from harmful content, maintaining honesty, offering personalized experiences, and the capacity for continuous improvement.

Claude v1's Limitations

Despite its advancements, Claude v1 encounters certain constraints:

- It relies solely on its training data for knowledge, limiting its understanding of many subjects.
- It is susceptible to occasional inaccuracies or irrelevant responses.
- Its expertise is confined to conversational tasks, lacking capabilities in data analysis or complex problem-solving.
- Claude v1 operates as a text-only interface and cannot directly interact with the physical world or process spoken language.

Falcon 40B

Falcon 40B is a member of the Falcon series of large language models (LLMs), developed by the Technology Innovation Institute (TII). This series also features models like Falcon 7B and Falcon 180B. Specifically, Falcon 40B is a causal decoder-only model tailored for a variety of natural language processing tasks.

This model boasts multilingual support, covering languages such as English, German, Spanish, and French, and offers partial proficiency in additional languages like Italian, Portuguese, Polish, Dutch, Romanian, Czech, and Swedish.

Model Design

The design of Falcon 40B (Figure 5-17) draws inspiration from the GPT-3 architecture but incorporates significant enhancements to boost its performance. It introduces rotary positional embeddings to improve its understanding of sequences. The model also benefits from advanced attention mechanisms, including multi-query attention and Flash Attention, alongside a decoder architecture that combines parallel attention with Multilayer Perceptron (MLP) structures, all under a dual-layer normalization framework to optimize computational efficiency.

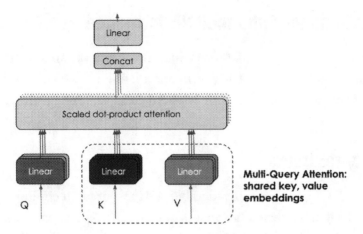

Figure 5-17. *Multi-query attention mechanism (Source: Falcon Blog)*

Data for Training

Falcon 40B's training involved over 1 trillion tokens derived from RefinedWeb, a web corpus meticulously filtered to ensure data quality, alongside other carefully selected datasets. TII focused on refining the data quality through extensive deduplication and rigorous filtering processes, ensuring a premium dataset that significantly contributes to the Falcon models' superior performance.

Training Process

The training of Falcon 40B utilized the power of AWS SageMaker equipped with 384 A100 40GB GPUs, adopting a 3D parallelism approach (Tensor Parallelism=8, Pipeline Parallelism=4, Data Parallelism=12) in harmony with ZeRO optimization. The training kicked off in December 2022 and was completed over a span of two months.

For enthusiasts and developers interested in training large language models using PyTorch, a comprehensive guide is available that walks through the entire process from setup to execution.

Multi-query Attention Mechanism

Falcon 40B introduces the multi-query attention mechanism, a departure from the conventional multi-head attention approach. Here, a single key and value pair is utilized across all attention heads, a strategy that, while not significantly altering pre-training outcomes, greatly enhances the model's scalability during inference.

CHAPTER 5   BASIC OVERVIEW OF THE COMPONENTS OF THE LLM ARCHITECTURES

Instruct Versions for Enhanced Performance

In addition to the standard model, TII has released instruct variants, such as Falcon-7B-Instruct and Falcon-40B-Instruct. These versions are fine-tuned with instructions and conversational data, optimizing them for assistant-like tasks and ensuring improved performance.

Accessibility for Users

Despite Falcon 40B's demand for considerable GPU resources, techniques like quantization make it more accessible on less powerful GPUs. For users with limited resources, the smaller Falcon 7B model is an alternative that can be utilized within platforms like Google Colab.

LLaMA

Meta AI initiated the LLaMA (Large Language Model Meta AI) project in February 2023, introducing a series of autoregressive large language models (LLMs) that marked a significant advancement in the field. The initial launch featured models of varying complexities, including those with 7, 13, 33, and 65 billion parameters.

Impressively, the 13-billion-parameter version outperformed the vastly larger GPT-3, which consists of 175 billion parameters, across numerous NLP benchmarks, while the largest LLaMA model showed competitive results against leading models like PaLM and Chinchilla.[7] Unlike previous high-capacity LLMs that were typically available through restricted APIs, Meta took an unprecedented step by making LLaMA's models openly accessible to researchers under a noncommercial license, although the model weights were leaked online shortly after their release.

In July 2023, Meta further expanded its LLaMA offerings with the introduction of LLaMA 2, developed in collaboration with Microsoft. This next iteration included models with 7, 13, and 70 billion parameters,[8] and although the architectural foundation remained

---

[7] Touvron, Hugo; Lavril, Thibaut; Izacard, Gautier; Martinet, Xavier; Lachaux, Marie-Anne; Lacroix, Timothée; Rozière, Baptiste; Goyal, Naman; Hambro, Eric; Azhar, Faisal; Rodriguez, Aurelien; Joulin, Armand; Grave, Edouard; Lample, Guillaume (2023). "LLaMA: Open and Efficient Foundation Language Models". arXiv:2302.13971 [cs.CL]

[8] "Meta and Microsoft Introduce the Next Generation of LLaMA". Meta. July 18, 2023. Retrieved July 21, 2023

similar to the first version, the data used for training was expanded by 40%. Meta's strategy involved not only releasing foundational models but also versions fine-tuned for dialog, dubbed LLaMA-2 Chat, which were made available for broad commercial use, albeit with certain restrictions that stirred debate regarding their open source status.

An evaluation by Patronus AI in November 2023 compared LLaMA 2 with other prominent AI models like GPT-4 and Anthropic's Claude 2 in a specialized test, revealing strengths and weaknesses in their abilities to process and interpret complex financial documents.

Leveraging the transformer architecture, LLaMA incorporates unique features such as the SwiGLU activation function,[9] rotary positional embeddings,[10] and root-mean-squared layer normalization[11] to enhance its performance, instead of standard layer normalization. The LLaMA 2 series further increased its context length capability, underscoring Meta's continuous efforts to push the boundaries of LLM efficiency and effectiveness.

Training for these models prioritized the augmentation of data volume over parameter count, with the LLaMA 1 models being trained on a dataset comprising 1.4 trillion tokens from diverse and publicly accessible sources. The LLaMA 2 models benefited from an even larger dataset, meticulously curated to enhance reliability and minimize privacy concerns, alongside specialized fine-tuning to optimize dialog interactions and ensure AI alignment through innovative training methods.

This evolution from LLaMA to LLaMA 2 not only showcases Meta's commitment to advancing AI technology but also emphasizes its intention to make powerful LLMs more accessible and applicable across a range of uses, setting a new standard in the development and deployment of language models.

LaMDA

Introduced by Google as the next iteration following Meena in 2020, LaMDA made its debut at the Google I/O keynote in 2021, with its advancement revealed in the subsequent year. This conversational AI model distinguishes itself by its capacity for engaging in unrestricted dialogues.

---

[9] Shazeer, Noam (2020-02-01), "GLU Variants Improve Transformer". arXiv:2104.09864 [cs.CL]
[10] Su, Jianlin; Lu, Yu; Pan, Shengfeng; Murtadha, Ahmed; Wen, Bo; Liu, Yunfeng (2021-04-01), "RoFormer: Enhanced Transformer with Rotary Position Embedding". arXiv:2104.09864 [cs.CL]
[11] Zhang, Biao; Sennrich, Rico (2019-10-01), "Root Mean Square Layer Normalization". arXiv:1910.07467 [cs.LG]

CHAPTER 5 BASIC OVERVIEW OF THE COMPONENTS OF THE LLM ARCHITECTURES

The development of LaMDA involved a comprehensive and detailed training regime, utilizing an enormous collection of documents, dialogues, and verbal exchanges that total over 1.56 trillion words. This extensive dataset facilitated LaMDA's understanding of diverse conversational dynamics.

Critical to refining LaMDA's performance were human evaluators, who assessed the model's outputs. Their assessments, grounded in search engine validations, were instrumental in enhancing LaMDA's precision and informational accuracy. These evaluations were based on the responses' usefulness, accuracy, and adherence to facts.

LaMDA's prowess emerges from its capacity to produce spontaneous conversations that aren't limited to specific tasks. It is adept at recognizing and adapting to various user intents, applying reinforcement learning techniques, and fluidly moving between topics without any apparent connection.

LaMDA builds upon **Google's BERT and transformer models**, with a specific emphasis on dialogue-driven tasks. It diverges from traditional NLP models that rely on written texts by training on a varied set of conversational data, including both text and audio recordings from real-life interactions.

A standout feature of LaMDA is its adeptness at generating responses tailored to the context of open-ended inquiries. It comprehends the underlying intent of user questions and crafts replies that reflect the ongoing conversation's context. For instance, LaMDA can tailor its responses about a particular movie by considering the user's past interactions and preferences alongside the broader conversational context.

Beyond dialogue systems, Google envisions LaMDA's utility extending to search and information retrieval, promising more pertinent, context-aware outcomes.

LaMDA signifies a leap forward in conversational AI, poised to markedly enhance human-machine interaction quality. Nevertheless, it raises concerns regarding privacy and the potential for unforeseen consequences, underscoring the need for vigilant oversight of its development and application.

LaMDA employs a transformer-based architecture tailored for dialogue modeling, consisting of an encoder for input processing and a decoder for response generation. The encoder utilizes self-attention layers to understand text relationships, capturing both immediate and overarching conversational contexts. The decoder, informed by the input, crafts context-specific replies through an additional attention mechanism focusing on relevant input text segments.

Incorporating features like multi-turn conversation handling and named entity recognition, LaMDA is engineered for richer, more relevant dialogue interactions. Its architecture facilitates nuanced conversations, setting a new standard for dialogue AI.

CHAPTER 5   BASIC OVERVIEW OF THE COMPONENTS OF THE LLM ARCHITECTURES

Google announced LaMDA 2 at its 2022 developer conference, showcasing an enhanced version of the AI system. Trained on extensive dialogue data, LaMDA 2 promises more natural, accurate, and safe conversations. Demonstrations highlighted its creative and on-topic response capabilities, alongside "List It," a feature for breaking down complex topics into actionable insights.

Google also introduced Bard, an experimental AI chat service, distinguishing itself from competitors by sourcing data directly from the Internet, promising fresh, high-quality answers. Based on the transformer architecture, Bard employs a streamlined version of LaMDA, aiming for widespread accessibility while leveraging Internet data for responses.

Guanaco-65B

Guanaco, a large language model (LLM), employs a fine-tuning technique known as LoRA, created by Tim Dettmers and colleagues within the University of Washington's Natural Language Processing group. Leveraging QLoRA, this approach enables the fine-tuning of models with as many as 65 billion parameters on a 48GB GPU, matching the performance of 16-bit models without degradation.

The Guanaco series of models surpasses the performance of all prior models on the Vicuna benchmark. However, due to their foundation on the LLaMA model family, their use in commercial settings is restricted.

QLoRA is a groundbreaking fine-tuning technique designed to significantly reduce memory requirements, enabling the fine-tuning of models with up to 65 billion parameters on a single 48GB GPU without compromising the quality of 16-bit fine-tuning tasks. This method innovatively directs gradient backpropagation through a statically quantized, 4-bit version of a pre-trained language model into Low Rank Adapters (LoRA).

The premier model suite, dubbed Guanaco, sets a new standard by outshining all previously available models on the Vicuna benchmark, achieving 99.3% of ChatGPT's performance with just 24 hours of fine-tuning on a solitary GPU. QLoRA's memory efficiency is achieved through several key innovations: the introduction of 4-bit NormalFloat (NF4), an optimally efficient data type for normally distributed weights; Double Quantization, which further reduces memory demands by quantizing quantization constants; and Paged Optimizers, designed to smooth out memory usage spikes.

Orca

Microsoft Research unveiled Orca 2, an advanced version of the LLaMA 2 language model, demonstrating performance on par with or surpassing models with ten times its parameter count. This leap in efficiency is attributed to a novel training approach involving a synthetic dataset and a technique known as Prompt Erasure.

In the development of Orca 2, a teacher-student learning framework is employed, where a larger, more adept language model (the teacher) guides a smaller, less complex model (the student) toward achieving performance levels akin to those of significantly larger counterparts.

This method allows the smaller model to learn various reasoning strategies and select the most suitable one for any given problem. The teacher model uses complex prompts to elicit specific reasoning behaviors, but with Prompt Erasure, these prompts are not passed to the student model. Instead, the student model receives only the task specifications and the expected outcome.

This approach enabled a 13-billion-parameter Orca 2 model to outshine a similarly sized LLaMA 2 model by 47.54% in benchmark tests. Moreover, the 7-billion-parameter version of Orca 2 was found to perform on a level "better or comparable" to a 70-billion-parameter LLaMA 2, particularly in reasoning tasks.

StableLM

Stability AI has introduced StableLM, marking a significant step forward in enhancing language comprehension within the realm of machine learning. The launch features two versions of StableLM, one equipped with 3 billion parameters and the other with 7 billion parameters.

The alpha release of StableLM invites users to explore and evaluate the model's performance, offering insights that will aid in refining its capabilities. Stability AI is gearing up to unveil two additional models, boasting 15 billion and 65 billion parameters, to further push the boundaries of language processing capabilities.

StableLM functions as a self-regressive model, adept at recognizing language patterns and crafting responses based on provided inputs. It comprises a foundational model that reliably predicts subsequent tokens and a more specialized model fine-tuned to adhere to explicit instructions. This fine-tuning process utilizes diverse datasets like Alpaca, GPT4All, Dolly, and HH, enhancing the model's proficiency in delivering customized responses and instructions.

The design of StableLM enables it to adjust to various contexts and linguistic styles, positioning it as a flexible asset for a wide array of language processing and generation tasks.

The unveiling of StableLM by Stability AI is noteworthy for its contribution to the natural language processing domain, particularly due to the model's transparent architecture and training approach. This openness sets it apart from competing models, encouraging the development and innovation community to expand and refine Stability AI's offerings, thereby promoting collaborative progress in the field.

Additionally, Stability AI is committed to broadening the accessibility of language models. By creating models capable of operating on consumer-grade hardware, such as smartphones and PCs, Stability AI is democratizing access to sophisticated language processing tools. This move not only widens the scope of potential applications but also empowers individuals and smaller entities to employ state-of-the-art language models for a variety of uses.

Palmyra

Palmyra Base underwent its primary training phase focusing on English language texts, although a small proportion of non-English content from Common Crawl was also included in its training dataset. Employing a causal language modeling (CLM) strategy for pretraining, Palmyra Base aligns with models like GPT-3 by incorporating only a decoder component in its architecture. This approach to training, centered on self-supervised causal language modeling, mirrors that of GPT-3, including the use of similar prompts and experimental frameworks for evaluation.

Palmyra Base distinguishes itself through its remarkable speed and capability, proving adept across a variety of sophisticated tasks, including sentiment analysis and content summarization.

The model, Palmyra Base (5b), was developed using a proprietary dataset curated by Writer.

Palmyra Base is engineered to internalize and represent the English language, making it a valuable tool for deriving features applicable to a range of downstream applications. Nonetheless, its primary strength lies in text generation based on prompts, the task for which it was originally designed and trained.

CHAPTER 5   BASIC OVERVIEW OF THE COMPONENTS OF THE LLM ARCHITECTURES

GPT4ALL

GPT4All emerges as a pioneering open source solution aimed at enhancing accessibility and privacy in the digital realm. Designed for users seeking a powerful, privacy-conscious chatBot that runs on local machines without the need for sophisticated hardware or Internet connectivity, GPT4All offers a blend of performance and privacy.

Key attributes of GPT4All include the following:

- **Locality and Cost-Free Operation:** It operates on local devices, eliminating the need for Internet access.

- **Compatibility with Consumer Hardware:** Engineered to function efficiently on standard CPUs, GPT4All is accessible without requiring high-end GPUs, though it does necessitate CPUs with AVX or AVX2 instruction set support.

- **Optimization for Performance:** Capable of handling language models ranging from 3 to 13 billion parameters on everyday computing devices like laptops and desktops.

- **Compact Model Sizes:** The models range between 3GB and 8GB, simplifying the download and integration process.

Ecosystem Components

- **GPT4All Backend:** The core of GPT4All, featuring a universally optimized C API designed for running extensive transformer decoder models. This API is adaptable to additional languages such as C++, Python, and Go, among others.

- **GPT4All Bindings:** Encapsulates the programming language bindings, inclusive of a Command-Line Interface (CLI).

- **GPT4All API:** Currently under development, this component will introduce REST API endpoints to facilitate the retrieval of language model completions and embeddings.

- **GPT4All Chat:** A cross-platform application available for macOS, Windows, and Linux users, offering chat functionality with added privacy and security through plug-ins like LocalDocs for interacting with local documents.

- **GPT4All Datasets:** Spearheaded by Nomic AI, this initiative provides Atlas, a platform for the streamlined management and curation of training datasets.

- **GPT4All Open Source Datalake:** Encourages community participation in enhancing the GPT4All model by sharing assistant tuning data in a privacy-conscious manner, where user consent is paramount.

The strength of GPT4All lies in its support for multiple model architectures, including the following:

- EleutherAI's GPT-J
- Meta's LLaMA
- Mosaic ML's MPT
- Replit
- TII's Falcon
- BigCode's StarCoder

These models are regularly updated to ensure they offer peak performance and quality, with GPT-J and MPT models, in particular, demonstrating impressive results compared to LLaMA, and ongoing innovations in MPT models suggesting exciting future enhancements.

Developed and maintained by Nomic AI, the GPT4All project stands as a testament to the potential for high-quality, secure, and locally operated chatbot solutions, offering versatility for both personal and professional use across various model architectures.

Summary

This chapter provides an overview of the components of large language model (LLM) architectures, essential for transforming raw textual data into meaningful, context-aware outputs. It begins with embedding layers, which convert tokens into continuous vector representations through an adaptable embedding matrix.

Next, feedforward neural networks (FFNs) are introduced, highlighting their role in processing input data through weighted sums and activation functions to recognize complex patterns. Recurrent layers are also discussed, emphasizing their importance in handling sequential data by maintaining hidden states and considering the context of earlier words.

The chapter covers also the attention mechanisms, such as self-attention and multi-head attention, crucial for focusing on relevant parts of input text. Transformers utilize these mechanisms to process extensive texts efficiently. Additionally, activation functions and normalization techniques are discussed for enhancing neural network performance and training stability.

In the next chapter, we explore diverse and impactful applications of large language models within the Python ecosystem like the following:

- **Text Generation:** Discusses methods and significance of generating human-like text for various content creation needs

- **Language Translation:** Highlights advancements and techniques in translating text across multiple languages with LLMs

- **Document Summarization:** Covers approaches to condense lengthy documents into concise summaries while retaining key information

- **Chatbots and Virtual Assistants:** Explains the construction and deployment of sophisticated AI-driven chatbots for customer support and other uses

- **Practical Applications:** Includes use cases such as customer support automation, efficient document search, and knowledge management

CHAPTER 6

Applications of LLMs in Python

In this chapter, we explore the diverse and impactful applications of large language models (LLMs) within the Python ecosystem. From enhancing natural language processing tasks to generating creative content, LLMs have become integral to many innovative solutions. We will delve into practical use cases, examine real-world examples, and provide insights into how these powerful models are transforming industries and driving technological advancements.

Text Generation and Creative Writing

Text generation stands as a sophisticated method that employs artificial intelligence and machine learning algorithms to produce text that closely mimics **human writing**. It empowers computers to generate coherent and contextually fitting text by discerning patterns and structures from existing textual data.

The Mechanism Behind Text Generation

Text generation algorithms primarily leverage **recurrent neural networks (RNNs) or transformer models** to grasp the intricate patterns and dependencies within a given text corpus. These models learn from the sequential arrangement of data, thereby enabling them to generate text that exhibits similar characteristics.

In the training phase, the algorithm is exposed to a vast amount of text data and learns to anticipate the subsequent word or sequence of words based on the context provided by preceding words. This iterative process entails refining the model's parameters to minimize the disparity between the predicted and actual text.

The Significance of Text Generation

Text generation holds significant importance across various domains, offering several advantages to businesses:

- **Content Creation:** By automating the creation of content such as blog posts, news articles, product descriptions, social media posts, and ad copy, text generation saves time and resources for businesses.

- **Chatbots and Virtual Assistants:** Text generation facilitates the development of chatbots and virtual assistants capable of interacting with users in a human-like manner, thereby delivering personalized responses and enhancing customer experiences.

- **Data Augmentation:** Text generation serves to augment training data for machine learning models, thereby enhancing their performance and generalization capabilities.

- **Personalization:** Through text generation techniques, personalized recommendations, advertisements, and messages tailored to individual users can be generated, thereby amplifying customer engagement and satisfaction.

Key Use Cases of Text Generation

Text generation finds application across diverse domains, including the following:

- **Natural Language Processing (NLP):** Text generation plays a pivotal role in various NLP tasks such as language translation, sentiment analysis, text summarization, and question answering.

- **Content Generation:** Text generation automates content creation for websites, blogs, marketing materials, and other media.

- **Virtual Assistants:** Virtual assistants powered by text generation provide conversational experiences and assist users in tasks such as scheduling, information retrieval, and personal assistance.

- **Data Augmentation:** Text generation is utilized to produce synthetic training data for machine learning models, particularly in scenarios where labeled data is limited.

What Is Creative Writing

Creative writing refers to the art of crafting original and imaginative works of literature or prose. It involves the expression of thoughts, ideas, and emotions in a unique and compelling manner, often using language in innovative and unconventional ways. Creative writing encompasses various forms such as poetry, short stories, novels, plays, screenplays, and essays.

Unlike technical or academic writing, which primarily focuses on conveying information or arguments in a clear and logical manner, creative writing emphasizes creativity, self-expression, and artistic freedom. Writers often draw inspiration from personal experiences, observations, and imagination to create vivid and engaging narratives that resonate with readers on an emotional level.

The process of creative writing can involve brainstorming, drafting, revising, and editing to refine ideas and bring them to life on the page. It requires careful attention to language, imagery, symbolism, and narrative structure to evoke specific moods, evoke emotions, and convey deeper meanings.

Utilizing LLMs for Creative Writing Endeavors

The advent of large language models (LLMs) has ushered in a realm of possibilities for creative writers and content creators alike. Let's delve into the myriad applications they offer.

1. Conceptualization and Brainstorming

Generating fresh and captivating ideas can often pose a challenge for writers. LLMs come to the rescue by furnishing lists of ideas, topics, or themes tailored to your specifications. From this array, you can cherry-pick the most intriguing concepts to further develop.

2. Composition and Refinement

In the drafting phase, LLMs serve as invaluable aids, crafting content based on your provided outlines or briefs. Furthermore, they assist in refining your work by suggesting alternate phrasings, rectifying grammatical errors, and enhancing the clarity and conciseness of your prose.

3. Dialogue Crafting and Characterization

The hallmark of any compelling narrative lies in its believable characters and dialogues. LLMs contribute to this aspect by generating authentic dialogues imbued with emotions, motivations, and distinct voices for each character. Moreover, they aid in fleshing out characters' backgrounds, descriptions, and other crucial facets.

4. World Building and Scene Setting

LLMs prove instrumental in fashioning immersive and intricately detailed settings for your narratives. By furnishing key parameters, these models conjure vivid descriptions and enriching world-building elements, spanning cultures, technologies, and historical contexts.

5. Poetry and Experimental Literature

Delving into the realm of poetry and experimental writing, LLMs offer avenues for exploration and innovation. Their ability to generate poetic verses or experiment with diverse writing styles provides fresh perspectives, serving as a springboard for crafting distinctive and imaginative pieces.

Blog Post Generator on a Topic and Length Provided by the User Based on OpenAI

> **Note** You need to install OpenAI with the command `pip install openai` and also to sign up on their website, pay for credit, if you don't set up automatic payments, and generate an API key for the utilization of the following app.

```
import openai

# Set up your OpenAI API key
openai.api_key = 'Your-API-Key'

def generate_blog_post(prompt):
    response = openai.chat.completions.create(
```

```python
        model="gpt-4",
        messages=[
    {
      "role": "system",
      "content": "You are an experienced blog post writer."
    },
    {
      "role": "user",
      "content": f"{prompt}"
    }
     ],
        max_tokens=1000,
        temperature=0.7,
        n=1,
        stop=None
    )
    return response.choices[0].message.content
def main():
    # Provide a prompt to start generating a blog post
    user_input = input("Enter your blog post topic: ")
    length = input("How many words?")
    prompt = f"""Write a blog post about the topic {user_input} and make it {length} words long."""

    # Generate the blog post
    blog_post = generate_blog_post(prompt)

    # Print the generated blog post
    print(blog_post)
if __name__ == "__main__":
    main()
```

Note Feel free to play with the temperature parameter of the model and you will see different nuances of the text generated.

As a result, the app will produce a blog post with a length close to the number of words you pointed to when the app prompted you to input it.

Language Translation and Multilingual LLMs

The integration of large language models (LLMs) and machine learning (ML) in the realm of language translation has yielded remarkable advancements in quality, precision, and efficiency. Unlike older translation systems, these models excel in comprehending and generating text akin to human-like expressions, thereby furnishing more refined and accurate translations. Their prowess is particularly evident in languages characterized by intricate patterns, homonyms, compound words, or where contextual nuances heavily influence meaning.

These strides are propelling us closer to a future where language barriers become obsolete. Illustrations from translations spanning Japanese, Chinese, German, French, and Italian to English serve as testaments to this evolution and the promising horizon of translation technology.

Language translation stands out as one of the marvels of AI and machine learning (ML), significantly augmenting global communication. With the integration of large language models (LLMs), translation quality, accuracy, and efficiency have undergone substantial enhancements.

LLMs, such as GPT-3, excel in comprehending and crafting text that mirrors human expression. Trained on extensive and diverse datasets, they exhibit an enhanced grasp of syntax, semantics, and the subtleties inherent in various languages. The outcome is a translation that is more nuanced and accurate, adept at tackling language intricacies that previously posed challenges.

Advantages of Utilizing LLMs for Translation

- **Contextual Understanding:** LLMs, having been trained on vast volumes of text data, possess the capability to capture nuanced meanings and contextual cues.
- **Enhanced Accuracy:** LLMs deliver significantly improved translation quality compared to traditional rule-based methods.

- **Multilingual Support:** LLMs are adept at facilitating translation across multiple languages, showcasing versatility and adaptability.

- **Continual Learning:** LLMs can be refined and updated with fresh data, thereby ensuring a continuous enhancement in translation quality.

How LLMs Translate Languages?

The process of training an LLM for text translation entails several stages. Initially, data collection and preprocessing are carried out. This typically involves assembling a sizable corpus of parallel text, comprising various versions of the same content in different languages. For instance, a parallel corpus might encompass English sentences alongside their corresponding translations in French, Spanish, or any other language.

Subsequently, the collected and preprocessed data are utilized to train the LLM. This training phase entails feeding the model pairs of sentences in the source language alongside their corresponding translations in the target language. The model analyzes the data patterns and relationships to acquire the skill of translating from the source language to the target language.

Challenges Associated with LLMs in Translation

- **Bias:** LLMs may exhibit biases, potentially resulting in inaccurate or offensive translations.

- **Cost:** The development and maintenance of LLMs can incur significant expenses.

- **Regulatory Considerations:** LLMs may be subject to regulatory oversight in certain jurisdictions.

While LLMs offer myriad benefits for translation, it's essential to address associated challenges. With the ongoing development of LLMs, efforts are underway to overcome these obstacles, paving the way for further advancements in translation technology. The emergence of large language models (LLMs) is poised to bring about significant transformations within the translation sector, presenting a multitude of potential impacts.

The Potential Impacts of LLMs on the Translation and Localization Industry

Enhanced Efficiency

LLMs boast the capability to translate text at a pace far exceeding that of human translators. This accelerated processing speed translates into invaluable time and cost savings for businesses, thereby streamlining workflows and enhancing operational efficiency.

Elevated Quality

Leveraging massive datasets encompassing extensive text and code, LLMs are adept at assimilating the intricacies inherent in various languages. This comprehensive training equips them to produce translations characterized by heightened accuracy and natural fluency. Consequently, businesses stand to benefit from translations that resonate authentically with target audiences, fostering improved communication and comprehension.

Pioneering Opportunities

The advent of LLMs heralds new prospects for translation endeavors, particularly in realms that have traditionally remained underserved or overlooked. These models can effectively navigate the translation of languages that were previously sidelined, including minority languages and specialized technical jargon. By facilitating the translation of content into such languages, LLMs open avenues to untapped markets, enabling businesses to extend their reach and diversify their clientele.

Translation App Based on the Google T5 Model

In order to use the app, you need to install the following packages before that in your notebook or terminal:

```
pip install sacremoses==0.0.53
pip install datasets==2.20.0
pip install transformers==4.41.2
pip install torch==2.3.0+cu121 torchvision==0.18.0+cu121 torchaudio==2.3.0+cu121
pip install sentencepiece==0.1.99
```

CHAPTER 6 APPLICATIONS OF LLMS IN PYTHON

Then execute the following code:

> **Note** Please note that the T5-Small model has a limitation in supporting all languages. Nonetheless, this guide will illustrate the procedure for utilizing translation models in our projects.

```
from datasets import load_dataset
from transformers import pipeline

t5_small_pipeline = pipeline(
    task="text2text-generation",
    model="t5-large",
    max_length=1000,
    model_kwargs={"cache_dir": '/content/Translation Test' },
)
text_to_translate = input("Enter your text for translation: ")
translate_from_language = input("Which language do you want to translate from: ")
translate_to_language = input("Enter your desired language to translate to: ")
prompt = f"translate from {translate_from_language} to {translate_to_language} - {text_to_translate}"

t5_small_pipeline(prompt)
```

Sample output:

Enter your text for translation: Hello, how is it going for you?
Which langauge do you want to translate from: english
Enter your desired language to translate to: german

[{'generated_text': 'Aus Deutsch in Englisch - Hallo, wie geht es für Sie?'}]

CHAPTER 6   APPLICATIONS OF LLMS IN PYTHON

Another example with OpenAI:

```python
import openai
openai.api_key = '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'

def translate_text(text, source_language, target_language):

    response = openai.chat.completions.create(
        model='gpt-4',
        messages=[
    {
      "role": "system",
      "content": "You are an experienced translator from one to another language."
    },
    {
      "role": "user",
      "content": f"Translate the following text from {source_language} to {target_language}:\n{text}"
    }
    ],
        max_tokens=100,
        n=1,
        stop=None,
        temperature=0.2,
        top_p=1.0,
        frequency_penalty=0.0,
        presence_penalty=0.0,
    )

    translation = response.choices[0].message.content
    return translation

# Get user input
text = input("Enter the text to translate: ")
source_language = input("Enter the source language: ")
target_language = input("Enter the target language: ")

translation = translate_text(text, source_language, target_language)
```

```
# print the translated text

print(f'Translated text: {translation}')
```

Output:

```
Enter the text to translate: hi how are you doing
Enter the source language: english
Enter the target language: german
Translated text: Hallo, wie geht es dir?
```

Text Summarization and Document Understanding

In the age of Big Data, the wealth of textual data highlights the critical significance of effective text summarization methods. Text summarization, which involves condensing lengthy documents or articles into clear and concise summaries while retaining their core meaning and essential information, is invaluable across diverse domains. Whether facilitating information retrieval or aiding content generation, summarization has become an indispensable element of natural language processing (NLP) applications.

Text summarization stands as a pivotal task within natural language processing (NLP), seeking to condense extensive text volumes into shorter yet coherent representations while retaining crucial information.

There exist three primary approaches to text summarization: *direct*, *abstractive*, and *extractive* summarization.

- **Direct Summarization:** This represents the most straightforward method, involving a direct request to the LLM to provide a summary of the document. While this approach is quick and uncomplicated, it might encounter difficulties with lengthy or intricate texts.

- **Abstractive Summarization:** This entails crafting a succinct summary that may include words, phrases, or sentences not directly present in the source text. This method hinges on contextual comprehension and the generation of human-like language to articulate the central ideas. Leveraging advanced language models like large language models (LLMs), abstractive summarization techniques aim to refine and rephrase content into a more concise form.

- **Extractive Summarization:** This aims to select and extract the most pertinent sentences or phrases directly from the source text to compose the summary. This approach does not involve rewriting or generating new sentences. Extractive summarization methods employ various techniques such as sentence scoring and ranking to pinpoint and extract the most significant content.

Article Summarization Application Using User-Provided URL

```
from langchain.output_parsers import PydanticOutputParser
from pydantic import field_validator
from pydantic import BaseModel, Field
from typing import List
from langchain.prompts import PromptTemplate
from langchain.llms import OpenAI
import requests
from newspaper import Article

openai_key = 'YOUR-API-KEY'

headers = {
    'User-Agent': 'Mozilla/5.0 (Windows NT 10.0; Win64; x64) AppleWebKit/537.36 (KHTML, like Gecko) Chrome/89.0.4389.82 Safari/537.36'
}

article_url = input("Enter Article URL To Be Summarized: ")

session = requests.Session()

try:
    response = session.get(article_url, headers=headers, timeout=10)

    if response.status_code == 200:
        article = Article(article_url)
        article.download()
        article.parse()
    else:
```

```
        print(f"Failed to fetch article at {article_url}")
except Exception as e:
    print(f"Error occurred while fetching article at {article_url}: {e}")

article_title = article.title
article_text = article.text

# create output parser class
class ArticleSummary(BaseModel):
    title: str = Field(description="Title of the article")
    summary: List[str] = Field(description="Bulleted list summary of the article")

    # validating whether the generated summary has at least three lines
    @field_validator('summary')
    def has_three_or_more_lines(cls, list_of_lines):
        if len(list_of_lines) < 3:
            raise ValueError("Generated summary has less than three bullet points!")
        return list_of_lines

# set up output parser
parser = PydanticOutputParser(pydantic_object=ArticleSummary)

# create prompt template
# notice that we are specifying the "partial_variables" parameter
template = """
You are an experienced content writing assistant that summarizes online articles.

Here's the article you want to summarize.

==================
Title: {article_title}

{article_text}
==================

{format_instructions}
"""
```

CHAPTER 6   APPLICATIONS OF LLMS IN PYTHON

```python
prompt = PromptTemplate(
    template=template,
    input_variables=["article_title", "article_text"],
    partial_variables={"format_instructions": parser.get_format_
    instructions()}
)

# Format the prompt using the article title and text obtained from scraping
formatted_prompt = prompt.format_prompt(article_title=article_title,
article_text=article_text)

# Instantiate model class
model = OpenAI(openai_api_key = openai_key, model_name="gpt-3.5-turbo-
instruct", temperature=0.0)

# Use the model to generate a summary
output = model(formatted_prompt.to_string())

# Parse the output into the Pydantic model
parsed_output = parser.parse(output.split("\"]}")[0] + "\"]}")
print("Title:", parsed_output.title)
print("Article Summary:")
for item in parsed_output.summary:
    print("-", item)
```

   Sample output:

```
Enter Article URL To Be Summarized: https://www.techtarget.com/whatis/
definition/large-language-model-LLM
Title: What are Large Language Models?
Article Summary:
- LLMs are becoming increasingly important to businesses as AI continues to
  grow and dominate the business setting.
- LLMs take a complex approach involving multiple components, including
  training on large volumes of data and using self-supervised learning and
  deep learning techniques.
```

- LLMs have a wide range of uses, including text generation, translation, content summary, rewriting, classification, sentiment analysis, and conversational AI.
- LLMs offer numerous advantages, such as extensibility, flexibility, performance, accuracy, and ease of training.
- However, there are also challenges and limitations to using LLMs, such as high development and operational costs, potential bias, lack of explainability, hallucination, complexity, and vulnerability to glitch tokens.

Explanation

This app is an extended version of the summarizing online articles app from the previous chapter using the OpenAI GPT-3.5 model. Before using it, you need to install the following packages:

- Langchain==0.1.4
- deeplake==3.9.11
- Openai==1.10.0
- Tiktoken==0.7.0
- Newspaper3k==0.2.8
- Pydantic==2.7.4

This is something you can do by executing the following command in your notebook or terminal: `pip install langchain==0.1.4 deeplake openai==1.10.0 tiktoken newspaper3k pydantic`. You also need an API key for OpenAI.

Here's a breakdown of what happens in the code:

1. It imports necessary modules and classes including "PydanticOutputParser", "BaseModel", "Field", "List", "PromptTemplate", "OpenAI", "requests", and "Article".

2. It prompts the user to input the URL of the article to be summarized.

3. It makes a request to fetch the article content from the provided URL and parses it using the "newspaper" library.

4. It defines a Pydantic model class called `"ArticleSummary"` with two fields: "title" and "summary". The "summary" field is defined with a validator "has_three_or_more_lines" ensuring that the generated summary has at least three bullet points.

5. It sets up an output parser using `"PydanticOutputParser"` for the `"ArticleSummary"` class.

6. It defines a template for the prompt using the article's title and text obtained from scraping.

7. It formats the prompt using the article title and text.

8. It instantiates the OpenAI model class with the specified API key and model name.

9. It uses the OpenAI model to generate a summary based on the formatted prompt.

10. It parses the output into the `"ArticleSummary"` Pydantic model.

11. It prints the parsed output, displaying the title and summary of the article.

Note OpenAI summarization

The actual summarization happens within the OpenAI model, where the input prompt is provided along with the article details, and the model generates a summary based on that. The Pydantic model is used to ensure the structure and validation of the output summary.

Question-Answering Systems: Knowledge at Your Fingertips

Enhancing Question-Answering Capabilities Through Large Language Models (LLMs)

The advent of large language models (LLMs) such as OpenAI's GPT has revolutionized the field of question answering (QA). These models elevate generative QA to new heights by facilitating a profound comprehension of both the query and its contextual nuances. LLMs possess the ability to craft interactions akin to human conversation, generating responses that align closely with the user's underlying intent. Moreover, they excel in providing intelligent and insightful summaries based on the underlying contexts they retrieve.

At its most basic level, a straightforward **generative QA system** requires only a user's textual query and an LLM. However, more sophisticated systems may undergo additional training on specialized domain knowledge. They may also integrate with search and recommendation engines to not only answer questions but also to seamlessly link relevant information sources. Through such advancements, these models transcend conventional search paradigms, evolving toward a more dynamic, responsive, and interactive mode of information retrieval.

Utilizing Large Language Models for Advanced Document Analysis

The integration of generative question answering models into information retrieval systems marks a significant advancement in document analysis. Leveraging large language models (LLMs) allows for the creation of more dynamic and intuitive systems capable of not only searching for data but also comprehending and interpreting it. This integration revolutionizes the processing of complex queries across diverse domains, fundamentally altering the way information is interacted with and retrieved.

CHAPTER 6 APPLICATIONS OF LLMS IN PYTHON

The Journey from Data to Response: A Comprehensive Overview

Document Parsing and Preparation

The process begins with loading and parsing documents of varied formats, such as text, PDFs, or database entries. Documents are segmented into manageable units, ranging from paragraphs to sentences, with the aid of natural language processing (NLP) packages like NLTK. These tools streamline tasks and handle intricacies like newlines and special characters, allowing engineers to focus on more advanced aspects.

Text Embedding and Indexing

Each textual unit undergoes conversion from character format to numerical vectors, generating embeddings that encapsulate semantic meaning. Models such as the Universal Sentence Encoder, DRAGON+, and Instructor are utilized to generate customized embeddings, either based on specific prompts or by leveraging the capabilities of large language models.

These embeddings are stored in a vector database, forming a searchable index for efficient information retrieval. Tools like NumPy, Faiss, or Elasticsearch/OpenSearch aid in this process, facilitating various algorithms for building indexes and managing retrieval tasks.

Query Processing and Context Retrieval

Upon user query submission, embedding techniques are applied to align with indexed data, facilitating context retrieval based on similarity metrics like cosine similarity.

The system retrieves the most pertinent textual units, establishing a contextual framework for generating answers.

Answer Generation

Functioning as a generative model, the LLM utilizes retrieved contexts alongside the query to formulate responses. By calculating conditional probabilities of word sequences, it generates contextually accurate and insightful answers.

This systematic approach underscores how generative QA (GQA) systems, augmented with LLMs, not only retrieve relevant information but also produce responses that deepen query understanding. These systems epitomize the next evolution in information retrieval, fostering interactive exchanges akin to conversing with a knowledgeable entity, delivering precise and insightful answers to specific inquiries.

Practical Applications and Use Cases of Generative Question Answering

Generative question answering (GQA) leverages advanced AI to provide detailed, context-aware responses to user queries, enhancing various real-world applications. GQA aids content creation, automating the generation of informative articles and reports. This versatility demonstrates GQA's transformative potential across multiple sectors, driving efficiency and innovation.

Enhanced Customer Support Through Automated Responses

Generative QA is reshaping customer support by enabling automated yet contextually aware responses. Leveraging LLMs, customer support systems deliver accurate answers tailored to individual query nuances, streamlining the support process with faster response times and reduced reliance on human agents for routine inquiries.

Efficient Search in Reports and Unstructured Documents

GQA is revolutionizing internal document search processes within organizations, particularly in handling complex documents like manufacturing reports, logistics records, and sales notes. Integration of vector databases enables efficient indexing and retrieval of relevant information, facilitating human-like interaction for employees. This approach ensures quick access to precise information, ranging from production data analysis to customer care insights.

Knowledge Management for Large Organizations

Generative QA systems provide significant benefits for large organizations with extensive knowledge repositories. Constructing comprehensive indexes of internal knowledge sources simplifies information retrieval, aiding in decision-making processes with insights drawn from the latest data. Whether querying company policies, historical records, or project reports, GQA systems offer an efficient means of managing and retrieving knowledge, streamlining organizational operations.

CHAPTER 6   APPLICATIONS OF LLMS IN PYTHON

Question Answering Chatbot over Documents with Sources

Let's delve into the realm of advanced artificial intelligence applications by constructing a sophisticated question answering (QA) chatbot tailored to operate over documents and furnish sources for its responses. Our QA chatbot harnesses a specialized mechanism known as the RetrievalQAWithSourcesChain, empowering it to meticulously navigate through a corpus of documents, discerning pertinent information to address inquiries.

This chain orchestrates structured prompts directed toward the underlying language model to elicit responses. These prompts are meticulously formulated to steer the language model's output, thereby enhancing the precision and relevance of the answers provided. Furthermore, the retrieval chain is intricately engineered to meticulously trace the origins of the information it retrieves, furnishing the capability to substantiate responses with credible references.

In our journey, we shall master the following techniques:

1. Scrape online articles provided by the user and archive the textual content along with the corresponding URLs.

2. Employ an embedding model to compute embeddings for these documents and archive them within Deep Lake, a vector database.

3. Segment the article texts into manageable fragments while meticulously noting the source of each segment.

4. Employ the RetrievalQAWithSourcesChain to craft a chatbot proficient in retrieving answers and concurrently tracking their sources.

5. Formulate responses to queries utilizing the chain, presenting the answer alongside its corroborative sources.

Before proceeding, ensure you have installed the necessary packages by executing the following command:

```
pip install langchain==0.1.4 deeplake==3.9.11 openai==0.27.8 tiktoken==0.7.0 newspaper3k==0.2.8
```

This command will install the required packages, including langchain version 0.1.4, deeplake, openai version 0.27.8, and tiktoken. Additionally, it will install newspaper3k package version 0.2.8.

Next, it's essential to incorporate your OpenAI and Deep Lake API keys into the environment variables. The LangChain library will then access these tokens and employ them for the integrations:

```
import os

os.environ["OPENAI_API_KEY"] = "<YOUR-OPENAI-API-KEY>"
os.environ["ACTIVELOOP_TOKEN"] = "<YOUR-ACTIVELOOP-API-KEY>"
```

Crawling the Articles Provided by the User

To initiate the process, let's gather some articles concerning AI news. Our focus lies in retrieving the textual content of each article along with its corresponding publication URL.

Within the code, you'll encounter the following components:

1. **Imports:** We commence by importing essential Python libraries. "`requests`" facilitates sending HTTP requests, "`newspaper`" proves invaluable for extracting and organizing articles from web pages, and "time" aids in incorporating pauses during our web scraping endeavor.

2. **Headers:** Certain websites may block requests lacking a proper User-Agent header, perceiving them as bot activity. Hence, we define a User-Agent string to emulate a genuine browser request.

3. **Article URLs:** A compilation of URLs for online articles.

4. **Web Scraping:** Establishing an HTTP session via "requests. Session()" enables us to execute multiple requests within the same session. Additionally, we initialize an empty list, "pages_content", designated for storing our scraped articles.

```
import requests
from newspaper import Article # https://github.com/codelucas/newspaper
import time
```

```python
headers = {
    'User-Agent': 'Mozilla/5.0 (Windows NT 10.0; Win64; x64)
    AppleWebKit/537.36 (KHTML, like Gecko) Chrome/89.0.4389.82
    Safari/537.36'
}

article_urls = [
    "https://www.site.com/2023/05/16/page-one/",
    "https://www.site.com/2023/05/16/page-two/",
    "https://www.site.com/2023/05/16/page-three/",
    Add Your URLs…
]

session = requests.Session()
pages_content = [] # where we save the scraped articles

for url in article_urls:
    try:
        time.sleep(2) # sleep two seconds for gentle scraping
        response = session.get(url, headers=headers, timeout=10)

        if response.status_code == 200:
            article = Article(url)
            article.download()
            article.parse()
            pages_content.append({ "url": url, "text": article.text })
        else:
            print(f"Failed to fetch article at {url}")
    except Exception as e:
        print(f"Error occurred while fetching article at {url}: {e}")
```

Following this, we'll proceed to compute the embeddings of our documents utilizing an embedding model and preserve them within Deep Lake, a database capable of handling multimodal vectors. OpenAIEmbeddings will serve as the tool for generating vector representations of our documents.

CHAPTER 6   APPLICATIONS OF LLMS IN PYTHON

These embeddings consist of high-dimensional vectors adept at encapsulating the semantic essence of the documents. Upon initializing an instance of the Deep Lake class, we furnish a path commencing with "hub://..." delineating the database name, which will subsequently be housed on the cloud.

```
from langchain.embeddings.openai import OpenAIEmbeddings
from langchain.vectorstores import DeepLake

embeddings = OpenAIEmbeddings(model="text-embedding-ada-002")

my_activeloop_org_id = "Your Organization ID - Usually Your Username"
my_activeloop_dataset_name = "qabot_with_source"
dataset_path = f"hub://{my_activeloop_org_id}/{my_activeloop_dataset_name}"

db = DeepLake(dataset_path=dataset_path, embedding=embeddings)
```

This segment is pivotal in configuring the system to handle the storage and retrieval of documents based on their semantic content. Such functionality stands as a linchpin for subsequent stages, where the objective is to pinpoint the most pertinent documents to address user queries.

Subsequently, we'll dissect these articles into smaller segments, with each segment's corresponding URL being preserved as a point of reference. This segmentation aids in streamlining data processing, rendering the retrieval task more manageable, and directing attention toward the most pertinent text fragments when responding to inquiries.

The RecursiveCharacterTextSplitter is instantiated with a chunk size of 1000 characters and an overlap of 100 characters between adjacent chunks. The "chunk_size" parameter delineates the length of each text segment, while "chunk_overlap" specifies the number of characters shared between adjacent segments. For every document within "pages_content", the text undergoes segmentation using the ".split_text()" method.

```
from langchain.text_splitter import RecursiveCharacterTextSplitter

text_splitter = RecursiveCharacterTextSplitter(chunk_size=1000, chunk_overlap=100)

all_texts, all_metadatas = [], []
for d in pages_content:
```

```
    chunks = text_splitter.split_text(d["text"])
    for chunk in chunks:
        all_texts.append(chunk)
        all_metadatas.append({ "source": d["url"] })
```

In the metadata dictionary, we utilize the "source" key to conform with the expectations of the RetrievalQAWithSourcesChain class, which autonomously retrieves this "source" item from the metadata. Subsequently, we incorporate these segmented chunks into our Deep Lake database alongside their corresponding metadata.

```
db.add_texts(all_texts, all_metadatas)
```

Let's dive into the exciting phase of constructing the QA Chatbot. We'll embark on developing a RetrievalQAWithSourcesChain, a chain designed not only to retrieve pertinent document excerpts for answering queries but also to maintain records of the sources associated with these documents.

Initiating the Chain Setup

To begin, we instantiate a RetrievalQAWithSourcesChain using the from_chain_type method. This method necessitates the following parameters:

1. LLM: This parameter anticipates an instance of a model, such as GPT-3, with a temperature of 0. The temperature parameter regulates the degree of randomness in the model's outputs – higher temperatures yield more randomness, whereas lower temperatures result in more deterministic outputs.

2. chain_type="stuff": This parameter delineates the type of chain utilized, influencing how the model processes retrieved documents and generates responses.

3. retriever=db.as_retriever(): This step establishes the retriever responsible for fetching relevant documents from the Deep Lake database. Here, the Deep Lake database instance, referred to as "db", undergoes conversion into a retriever utilizing its as_retriever method.

```
from langchain.chains import RetrievalQAWithSourcesChain
from langchain import OpenAI

llm = OpenAI(model_name="gpt-3.5-turbo-instruct",
temperature=0)

chain = RetrievalQAWithSourcesChain.from_chain_type(llm=llm,
                                  chain_type="stuff",
                                  retriever=db.as_retriever())
```

Finally, we'll utilize the chain to generate a response to a question. This response will encompass both the answer to the question and its associated sources.

```
d_response = chain({"question": "What does Geoffrey Hinton think about recent trends in AI?"})

print("Response:")
print(d_response["answer"])
print("Sources:")
for source in d_response["sources"].split(", "):
    print("- " + source)
```

Full Code of the App

```
import requests
from newspaper import Article # https://github.com/codelucas/newspaper
import time

headers = {
    'User-Agent': 'Mozilla/5.0 (Windows NT 10.0; Win64; x64) AppleWebKit/537.36 (KHTML, like Gecko) Chrome/89.0.4389.82 Safari/537.36'
}

article_urls = [
    "https://www.site.com/2023/05/16/page-one/",
    "https://www.site.com/2023/05/16/page-two/",
    "https://www.site.com/2023/05/16/page-three/",
    Add Your URLs…
]
```

```python
session = requests.Session()
pages_content = []

for url in article_urls:
    try:
        time.sleep(2) # sleep two seconds for gentle scraping
        response = session.get(url, headers=headers, timeout=10)

        if response.status_code == 200:
            article = Article(url)
            article.download()
            article.parse()
            pages_content.append({ "url": url, "text": article.text })
        else:
            print(f"Failed to fetch article at {url}")
    except Exception as e:
        print(f"Error occurred while fetching article at {url}: {e}")

from langchain.embeddings.openai import OpenAIEmbeddings
from langchain.vectorstores import DeepLake

embeddings = OpenAIEmbeddings(model="text-embedding-ada-002")

my_activeloop_org_id = "Your Username"
my_activeloop_dataset_name = "langchain_course_qabot_with_source"
dataset_path = f"hub://{my_activeloop_org_id}/{my_activeloop_dataset_name}"

db = DeepLake(dataset_path=dataset_path, embedding=embeddings)

from langchain.text_splitter import RecursiveCharacterTextSplitter

text_splitter = RecursiveCharacterTextSplitter(chunk_size=1000, chunk_overlap=100)

all_texts, all_metadatas = [], []
for d in pages_content:
    chunks = text_splitter.split_text(d["text"])
    for chunk in chunks:
        all_texts.append(chunk)
        all_metadatas.append({ "source": d["url"] })
```

```
db.add_texts(all_texts, all_metadatas)

from langchain.chains import RetrievalQAWithSourcesChain
from langchain import OpenAI

llm = OpenAI(model_name="gpt-3.5-turbo-instruct", temperature=0)

chain = RetrievalQAWithSourcesChain.from_chain_type(llm=llm,
                                                    chain_type="stuff",
                                                    retriever=db.as_
                                                    retriever())

d_response = chain({"question": "Your Question?"})

print("Response:")
print(d_response["answer"])
print("Sources:")
for source in d_response["sources"].split(", "):
    print("- " + source)
```

Example output:

Should I add nofollow to my paginated pages?
Response:
It is recommended to use the same titles and descriptions for all pages in a paginated sequence and to give each page a unique URL. It is also advised to avoid using the first page of a paginated sequence as the canonical page and to avoid indexing URLs with filters or alternative sort orders. It is not necessary to add nofollow to paginated pages.

Sources:
https://developers.google.com/search/docs/specialty/ecommerce/pagination-and-incremental-page-loading

CHAPTER 6   APPLICATIONS OF LLMS IN PYTHON

Chatbots and Virtual Assistants

The advent of sophisticated language models such as GPT-4 has revolutionized the design of chatbots, pushing boundaries further than ever before. Understanding the diverse categories of chatbots, the methodologies driving their design, and adhering to best practices are essential for crafting an AI-driven assistant that truly delivers.

Education is experiencing a continual transformation, with technology assuming a pivotal role in driving progress. A cutting-edge trend in educational technology involves the integration of chatbots into Master of Laws (LLM) programs. These digital assistants are adept at promptly addressing user inquiries and facilitating immediate access to information, showcasing their potential to enhance learning experiences.

What Is the Concept Behind Chatbots?

Chatbots represent AI-driven solutions designed to emulate human interactions while transparently operating as automated services. They facilitate real-time engagement between businesses and customers, addressing inquiries, completing tasks, and processing transactions. By doing so, they enable customer service teams to focus on resolving more intricate issues.

Originating as far back as the 1960s, chatbots boast a rich history, with notable predecessors including ELIZA and Siri. Their functionality relies on diverse approaches, ranging from predefined templates and keyword recognition to sophisticated techniques like natural language processing (NLP) and machine learning (ML), enabling them to comprehend and respond to user input effectively.

Chatbots excel at handling a wide array of tasks, from providing customer support to efficiently executing repetitive tasks on a large scale. Moreover, they seamlessly facilitate the transition of conversations to human agents when required. Offering a versatile omni-channel experience, chatbots operate across various platforms, including websites, messaging applications, virtual assistants, and even traditional telephone systems.

Practical Applications of LLM-Trained Chatbots

The utilization of LLMs allows for the training of chatbots tailored to specific functions, opening up a spectrum of deployment opportunities across various departments.

- **Sales and Marketing:** Leveraging LLM-trained chatbots can streamline sales processes by guiding prospects, offering bespoke recommendations, and crafting detailed customer profiles, thereby amplifying sales and marketing endeavors.

- **Content Marketing:** These chatbots equipped with LLM knowledge possess the ability to curate customized content, deliver tailored recommendations, automate content dissemination, and solicit feedback from customers, bolstering content marketing strategies.

- **Customer Support:** Empowered by LLM capabilities, chatbots excel in autonomously managing customer inquiries, ensuring consistent and precise responses, addressing frequently asked questions, and furnishing real-time information to enrich customer support interactions.

- **Social Media Marketing and Lead Generation:** Chatbots imbued with insights from LLMs can efficiently collect user data, engage in social media dialogues, and keep customers abreast of the latest releases, events, and promotions, thereby amplifying social media marketing initiatives and lead generation efforts.

- **Training and Development:** Integration of LLM-trained chatbots into training regimens guarantees uninterrupted access to learning materials, personalized learning pathways, round-the-clock resolution of queries, and instantaneous feedback for learners, revolutionizing training and development strategies.

Guide to Building a Chatbot with LLMs

Building a chatbot with large language models (LLMs) offers a pathway to creating sophisticated, responsive, and intelligent conversational agents. Key aspects include model selection and model fine-tuning, data preprocessing, integration and deployment (deployment is out of the scope of this book), and ensuring privacy and security. By following these guidelines, developers can create chatbots that deliver high-quality user experiences, effectively automate customer service, and enhance engagement across various applications.

Model Selection

The initial step is crucial – selecting the right language model. Choices abound, from the likes of GPT-3, GPT-Neo, GPT-2, to BERT, each differing in size, capabilities, and pre-trained weights. Opt for a model that harmonizes with the chatbot's intended purpose and available resources.

Data Preprocessing and Cleansing

Prepare the dataset meticulously for training the chatbot. Weed out noise and extraneous information from the conversational data. Data preprocessing ensures the dataset is refined, facilitating the model to learn and generate coherent responses effectively.

Fine-Tuning the Model

Next comes fine-tuning the selected language model on the pristine conversational dataset. This involves tweaking the model's parameters to suit the specific conversational context, thereby enhancing its response generation capabilities.

Integration and Deployment

Integrate the fine-tuned model seamlessly into a chatbot framework or platform. Leverage open source frameworks like Hugging Face Transformers, TensorFlow, or PyTorch for efficient model integration. Deploy the chatbot across diverse platforms such as websites, messaging applications, or APIs for widespread accessibility.

Best Practices and Considerations

- **Performance Evaluation:** Regularly evaluate the chatbot's performance to ensure its responses remain pertinent to user queries and align with the intended context.
- **Privacy and Security Measures:** Implement robust security protocols to safeguard user data, particularly when handling sensitive information, thus fostering trust and confidentiality.

- **Continuous Learning and Enhancement:** Enable the chatbot to learn from user interactions, dynamically adapting and refining its responses based on evolving conversational patterns, thereby ensuring continual improvement in user experience.

Customer Support Question Answering Chatbot

Initially, we gather content from online articles, segmenting them into smaller sections, then calculating their embeddings for storage in Deep Lake. Subsequently, we employ a user query to extract the most pertinent sections from Deep Lake, assembling them into a prompt for the LLM to generate the final response.

It's crucial to acknowledge the potential for generating hallucinations or misinformation when utilizing LLMs. While this might not align with the standards for many customer support scenarios, the chatbot can still aid operators in drafting responses that they can verify before sending to users.

Moving forward, we'll delve into managing conversations with GPT-3 and present examples to illustrate the efficacy of this workflow.

Firstly, ensure the OPENAI_API_KEY and ACTIVELOOP_TOKEN environment variables are configured with your respective API keys and tokens.

Since we'll be utilizing the SeleniumURLLoader LangChain class, which relies on the unstructured and selenium Python libraries, let's install it via pip. It's advisable to install the latest version of the library. However, please note that the code has been specifically tested on version 0.7.7.

```
pip install unstructured==0.14.9 selenium==4.22.0
```

Ensure you've installed the required packages by running the following command: "pip install langchain==0.0.208 deeplake openai==0.27.8 tiktoken". Now, let's proceed to import the necessary libraries.

```
from langchain.embeddings.openai import OpenAIEmbeddings
from langchain.vectorstores import DeepLake
from langchain.text_splitter import CharacterTextSplitter
from langchain import OpenAI
from langchain.document_loaders import SeleniumURLLoader
from langchain import PromptTemplate
```

These libraries offer capabilities for managing OpenAI embeddings, handling vector storage, text segmentation, and interfacing with the OpenAI API. They facilitate the development of a context-aware question-answering system, integrating retrieval and text generation functionalities. Our chatbot's database will primarily comprise articles concerning technical issues.

```
urls = ['<YOUR URL 1>', '<YOUR URL 2>']
```

Step 1: Document Segmentation and Embedding Calculation

We retrieve documents from the provided URLs and segment them into chunks using the CharacterTextSplitter with a chunk size of 1000 and no overlap.

```
loader = SeleniumURLLoader(urls=urls)
docs_not_splitted = loader.load()

text_splitter = CharacterTextSplitter(chunk_size=1000, chunk_overlap=0)
docs = text_splitter.split_documents(docs_not_splitted)
```

Subsequently, we compute the embeddings using OpenAIEmbeddings and save them in a Deep Lake vector store on the cloud. In an optimal production setup, we could upload entire websites or course lessons to a Deep Lake dataset, enabling searches across thousands or even millions of documents. Leveraging a cloud-based serverless Deep Lake dataset ensures that applications operating from various locations can effortlessly access the same centralized dataset without necessitating the deployment of a vector store on a bespoke machine.

Prior to running the following code, ensure your OpenAI key is stored in the "OPENAI_API_KEY" environment variable.

```
embeddings = OpenAIEmbeddings(model="text-embedding-ada-002")

my_activeloop_org_id = "Your Username"
my_activeloop_dataset_name = "langchain_course_customer_support"
dataset_path = f"hub://{my_activeloop_org_id}/{my_activeloop_dataset_name}"
db = DeepLake(dataset_path=dataset_path, embedding=embeddings)

# add documents to our Deep Lake dataset
db.add_documents(docs)
```

Step 2: Formulate a Prompt for GPT-3 Utilizing Recommended Techniques

We will construct a prompt template that integrates role-prompting, pertinent Knowledge Base data, and the user's inquiry.

```
template = """You are an exceptional customer support chatbot that gently answer questions.

You know the following context information.

{chunks_formatted}

Answer to the following question from a customer. Use only information from the previous context information. Do not invent stuff.

Question: {query}

Answer:"""

prompt = PromptTemplate(
    input_variables=["chunks_formatted", "query"],
    template=template,
)
```

The framework establishes the chatbot's persona as an exceptional customer support assistant. It takes two input variables: "chunks_formatted," comprising pre-formatted segments from articles, and "query," representing the customer's inquiry. The goal is to produce an accurate response using only the provided segments, avoiding the generation of false or fabricated information.

Step 3: Employ the GPT-3 Model with a Temperature of 0 for Text Generation

To generate a response, we initially retrieve the top-k segments most similar to the user query. We then format the prompt and submit it to the GPT-3 model with a temperature of 0.

```
# user question
```

CHAPTER 6   APPLICATIONS OF LLMS IN PYTHON

```
query = "Your Query"

# retrieve relevant chunks
docs = db.similarity_search(query)
retrieved_chunks = [doc.page_content for doc in docs]

# format the prompt
chunks_formatted = "\n\n".join(retrieved_chunks)
prompt_formatted = prompt.format(chunks_formatted=chunks_formatted,
query=query)

# generate answer
llm = OpenAI(model="gpt-3.5-turbo-instruct", temperature=0)
answer = llm(prompt_formatted)
print(answer)
```

The code of the whole application:

```
from langchain.embeddings.openai import OpenAIEmbeddings
from langchain.vectorstores import DeepLake
from langchain.text_splitter import CharacterTextSplitter
from langchain import OpenAI
from langchain.document_loaders import SeleniumURLLoader
from langchain import PromptTemplate

urls = ['<YOUR URL1>', '<YOUR URL 2>']

# use the selenium scraper to load the documents
loader = SeleniumURLLoader(urls=urls)
docs_not_splitted = loader.load()

# we split the documents into smaller chunks
text_splitter = CharacterTextSplitter(chunk_size=1000, chunk_overlap=0)
docs = text_splitter.split_documents(docs_not_splitted)

# Before executing the following code, make sure to have
# your OpenAI key is saved in the "OPENAI_API_KEY" environment variable.
embeddings = OpenAIEmbeddings(model="text-embedding-ada-002")

# create Deep Lake dataset
```

```python
# TODO: use your organization id here. (by default, org id is your
username)
my_activeloop_org_id = "YOUR USERNAME"
my_activeloop_dataset_name = "customer_support"
dataset_path = f"hub://{my_activeloop_org_id}/{my_activeloop_dataset_name}"
db = DeepLake(dataset_path=dataset_path, embedding=embeddings)

# add documents to our Deep Lake dataset
db.add_documents(docs)

# let's write a prompt for a customer support chatbot that
# answer questions using information extracted from our db
template = """You are an experienced customer support chatbot that answers questions in a comprehensive way.

You know the following context information.

{chunks_formatted}

Answer to the following question from a user. Use only information from the previous context information. Do not be creative.

Question: {query}

Answer:"""

prompt = PromptTemplate(
    input_variables=["chunks_formatted", "query"],
    template=template,
)

# the full pipeline

# user question
query = "Your question?"

# retrieve relevant chunks
docs = db.similarity_search(query)
retrieved_chunks = [doc.page_content for doc in docs]

# format the prompt
```

```python
chunks_formatted = "\n\n".join(retrieved_chunks)
prompt_formatted = prompt.format(chunks_formatted=chunks_formatted,
query=query)

# generate answer
llm = OpenAI(model="gpt-3.5-turbo-instruct", temperature=0)
answer = llm(prompt_formatted)
print(answer)
```

Example output:

```
Deep Lake Dataset in hub://didogrigorov/test_vector_db already exists,
loading from the storage
Creating 38 embeddings in 1 batches of size 38::
100%|████████████████████| 1/1 [00:11<00:00, 11.84s/it]
Dataset(path='hub://didogrigorov/test_vector_db', tensors=['embedding',
'id', 'metadata', 'text'])
```

tensor	htype	shape	dtype	compression
embedding	embedding	(58, 1536)	float32	None
id	text	(58, 1)	str	None
metadata	json	(58, 1)	str	None
text	text	(58, 1)	str	None

AI-powered scams are fraudulent activities that use artificial intelligence technology to make them more convincing, easier, or cheaper to execute. These scams can take various forms, such as voice cloning of family and friends, fake identities and verification fraud, and even fake media generated by AI. It is important to be vigilant and cautious when receiving emails or messages from unknown sources, and to never click on suspicious links or open attachments unless you are 100% sure of their authenticity.

Basic Prompting – The Common Thing Between All Applications Presented

Prompting lies at the heart of engaging with large language models (LLMs) and guiding them to produce desired outputs. Here's an exploration into fundamental prompting and some useful templates to kickstart your journey.

Understanding Prompting

Prompting involves furnishing instructions or cues to an LLM, informing it of the desired task. This could be a simple query, an elaborate task description, or even a creative stimulus. Clarity, brevity, and specificity in instructions are crucial for optimal outcomes.

Benefits of Effective Prompting

- **Enhanced Accuracy:** Clear prompts enable LLMs to comprehend intent better, resulting in more relevant and precise outputs.

- **Stimulated Creativity:** By providing specific details or constraints, you can steer LLMs toward generating innovative and unique textual compositions.

- **Improved Efficiency:** Well-crafted prompts streamline the process, ensuring LLMs focus on the intended task promptly, saving time and effort.

Fundamental Prompting Techniques

1. **Start with Simplicity:** Initiate with straightforward prompts, such as questions or explicit directives.

2. **Precision Is Key:** Incorporate keywords, examples, and pertinent details to guide LLMs toward the desired outcome effectively.

3. **Consider Tone and Style:** Specify the desired tone (formal, informal, humorous) and style (poem, script, email) for tailored results.

4. **Utilize References:** Provide examples or reference materials to offer LLMs a clearer understanding of expectations.

5. **Experiment and Iterate:** Embrace experimentation, refining prompts based on outcomes to optimize performance.

Prompt Template Examples

Information Retrieval

- **Question Answering:** "What is the capital of Spain?"
- **Summarization:** "Summarize the key moments of the following research paper."
- **Topic Exploration:** "Provide interesting facts about…"

Creative Writing

- **Story Starter:** "Once upon a time…"
- **Poem Generator:** "Compose a poem in the style of … on the topic …."
- **Script Writing:** "Create a movie scene showcasing a romantic conversation between two characters."

Code Generation

- **Function Creation:** "Write a Python function that takes two numbers as input and returns their sum."
- **Code Translation:** "Translate this C++ code into Python."
- **Bug Fixing:** "Fix the syntax error in the following code snippet."

Translation

- **Language Translation:** "Translate this sentence from English to German."
- **Dialect Conversion:** "Rewrite this text in American English."
- **Formal/Informal Conversion:** "Write a more formal version of this email."
- **Grammatical Errors:** "Is this text grammatically correct?"

Summary

This chapter delves into the diverse and impactful applications of large language models (LLMs) within the Python ecosystem, highlighting their ability to enhance natural language processing tasks and generate creative content. The chapter covers key use cases such as text generation, language translation, and document summarization, explaining the underlying mechanisms and practical benefits in each domain.

It also explores the construction of advanced chatbots and virtual assistants using LLMs, providing insights into model selection, data preprocessing, and integration strategies. Emphasis is placed on practical applications, including customer support automation, efficient search in unstructured documents, and knowledge management in large organizations. The chapter concludes by discussing best practices for building effective and secure LLM-driven solutions, ensuring continuous learning and improvement in user experience.

In the upcoming final chapter, we will explore how to build real-life applications using LLMs with

- LangChain
- Hugging Face
- Pinecone
- OpenAI
- Cohere
- Lamini.ai

CHAPTER 7

Harnessing Python 3.11 and Python Libraries for LLM Development

In the ever-evolving landscape of artificial intelligence, large language models (LLMs) have emerged as powerful tools for a variety of applications, from natural language processing to sophisticated AI-driven solutions. The advent of Python 3.11 brings a host of new features and optimizations that significantly enhance the development of these complex models. Coupled with an array of robust Python libraries, this chapter delves into the practical aspects of leveraging Python 3.11 to develop LLMs-based applications effectively with platforms like LangChain, Hugging Face, Pinecone, OpenAI, Cohere, and Lamini.ai.

We explore the various types of data sources and the importance of data collection strategies. From there, we will guide you through the processes of data cleaning, normalization, and transformation, highlighting the significance of each step in eliminating noise and enhancing data quality. Special attention will be given to tokenization and embedding techniques, which are crucial for converting textual data into a form that LLMs can effectively process and understand.

LangChain

LangChain is a public, open source platform designed to empower developers working in the realm of artificial intelligence (AI) and machine learning. It facilitates the integration of expansive language models with various external systems, thereby enabling the creation of applications powered by large language models (LLMs). LangChain's primary objective is to forge connections between robust LLMs, such as OpenAI's GPT-3.5 and GPT-4, Cohere, and multiple external data sources. This integration aims to enhance the development and utilization of natural language processing (NLP) applications.

The framework is accessible to developers, software engineers, and data scientists proficient in Python, JavaScript, or TypeScript, providing packages in these languages. LangChain was initiated as a public, open source endeavor by Harrison Chase and Ankush Gola in 2022, with its first version also being released within the same year.

The significance of LangChain lies in its ability to streamline the creation of generative AI applications. It offers a simplified avenue for developers to build sophisticated NLP applications by organizing and making accessible large volumes of data. This is particularly beneficial for LLMs that need to process and access vast datasets.

LangChain Features

LangChain encompasses a suite of components designed to enhance the development and functionality of NLP applications:

- **Model Interaction:** This facet, also known as model input/output (I/O), facilitates engagement with any language model. It handles the administration of input to models and the interpretation of output data.

- **Data Connection and Retrieval:** This functionality allows for the transformation, storage, and retrieval of data accessible to LLMs. Data can be stored in databases and fetched via queries.

- **Chains:** For the creation of intricate applications, LangChain enables the integration of various components or multiple LLMs. This process creates what are known as LLM chains, connecting different models and tools.

- **Agents:** Through the agent module, LLMs can determine optimal actions for problem-solving. This is achieved by directing a sequence of commands to LLMs and other resources, guiding them to fulfill specific requests.

- **Memory:** This component aids LLMs in retaining the context of user interactions. It allows for the incorporation of both short-term and long-term memory, tailored to the application's requirements.

What Are the Integrations of LangChain?

LangChain also supports applications through integrations with LLM providers and external data sources. It can create chatbots or question-answering systems by merging LLMs from entities like Hugging Face, Cohere, and OpenAI with data repositories such as Apify Actors, Google Search, and Wikipedia. This fusion allows applications to process user queries and source optimal responses from these platforms.

Additionally, LangChain can integrate with cloud storage services like Amazon Web Services, Google Cloud, and Microsoft Azure, and vector databases for storing and querying high-dimensional data, such as Pinecone. These integrations leverage cutting-edge NLP technologies to craft effective and efficient applications.

How to Build Applications in LangChain?

Creating applications with LangChain involves leveraging language model capabilities to build customized apps.

The development process generally follows several essential steps:

1. **Define the Application's Purpose:** Initially, the developer needs to establish the application's specific function and scope. This includes identifying the necessary integrations, components, and language models that will be used.

2. **Develop the Application Logic:** Using prompts, developers can construct the logic or functionality that the application will follow.

3. **Customization of Functionality:** LangChain provides the flexibility for developers to adjust and modify its code, enabling the creation of tailored functionalities that cater to the specific requirements of the application.

4. **Optimization of Language Models:** Selecting the right language model for the task and tuning it to match the application's specific needs is crucial for optimal performance.

5. **Data Preparation:** Ensuring the cleanliness and accuracy of data through cleansing techniques is vital, along with implementing security protocols to safeguard sensitive information.

6. **Continuous Testing:** To maintain the application's efficiency and reliability, ongoing testing is necessary.

This approach enables the development of robust applications that harness the power of language models to meet diverse and specific use cases.

Use Cases of LangChain

LangChain's capabilities in harnessing large language models (LLMs) unlock a vast array of sophisticated applications across numerous sectors and industries. Here are some illustrative examples and use cases:

- **Customer Support Chatbots:** LangChain facilitates the creation of advanced chatbots that can manage intricate inquiries and even conduct transactions. These bots are designed to grasp and remember the context of a user's conversation, similar to how ChatGPT operates, enhancing customer service and experience.

- **Coding Assistants:** Leveraging LangChain, alongside APIs from platforms like OpenAI, enables the development of tools that aid software developers and tech professionals in refining their coding capabilities and boosting productivity.

- **Healthcare Innovations:** In the healthcare domain, applications built with LangChain are revolutionizing the way diagnoses are made and streamlining administrative tasks such as scheduling appointments. This automation allows healthcare professionals to dedicate more time to critical patient care.

- **Marketing and E-commerce Tools:** LLM-powered applications are transforming e-commerce and marketing by understanding consumer behavior, purchasing patterns, and product details. This enables the generation of personalized product recommendations and engaging product descriptions, helping businesses attract and retain customers.

These examples underscore LangChain's versatility in creating solutions that address complex needs across a wide range of fields, from improving customer interactions to supporting healthcare providers and enhancing e-commerce strategies.

Example of a LangChain App – Article Summarizer

Workflow

- **Install Essential Packages:** Begin with installing the required packages: requests, newspaper3k, and langchain.

- **Data Collection:** Use the requests package to retrieve content from specific article URL.

- **Extract Information:** Leverage the newspaper package to parse the collected HTML, isolating article titles and body text.

- **Preprocess Text:** Clean and structure the extracted content to prepare it for input into ChatGPT.

- **Generate Summaries:** Employ ChatGPT to produce concise summaries of the article' content.

- **Display Results:** Present the summaries alongside the original titles, offering a quick insight into each article's main points.

This application utilizing ChatGPT enables you to quickly grasp essential information from articles with the help of AI summarization. It is designed to keep you well informed without dedicating extensive time to reading full articles, showcasing the practicality of AI in enhancing information consumption efficiency.

Start by securing your OpenAI API key, a prerequisite for utilizing the summarizer. This involves creating an account on the OpenAI website and gaining API access. Once your account is set up, locate the API keys section to retrieve your key.

Ensure the installation of necessary libraries by executing the following command: `pip install langchain==0.1.4 deeplake openai==1.10.0 tiktoken`. It's also important to install the newspaper3k library, specifically version **0.2.8**, as it is the version verified for compatibility in this tutorial.

In your Python script or notebook, assign your API key to an environment variable named "OPENAI_API_KEY". To accomplish this using a ".env" file, employ the "load_dotenv" function.

CHAPTER 7   HARNESSING PYTHON 3.11 AND PYTHON LIBRARIES FOR LLM DEVELOPMENT

In the following app:

1. You need to select an article's URL to create a summary. Add it to the variable with value "YOUR-URL".

2. The subsequent script retrieves the article from a set of URLs utilizing the requests library alongside a personalized User-Agent header.

3. It subsequently isolates the title and content of each article using the newspaper library.

4. You need to install Python Dotenv using the command `pip install python-dotenv`.

5. To generate a .env file, go to your project directory using the terminal and execute the touch command in the following manner: `touch .env`.

6. Add there your API key in the form of a `variable = OPEN_AI_KEY = Your API Key`.

7. Then load the environmental variables:

   ```
   from dotenv import load_dotenv
   load_dotenv()
   ```

Then use the following code:

```
import json
from dotenv import load_dotenv
load_dotenv()

import requests
from newspaper import Article

headers = {
    'User-Agent': 'Mozilla/5.0 (Windows NT 10.0; Win64; x64)
AppleWebKit/537.36 (KHTML, like Gecko) Chrome/89.0.4389.82 Safari/537.36'
}

article_url = "YOUR-URL"
```

```
session = requests.Session()

try:
    response = session.get(article_url, headers=headers, timeout=10)

    if response.status_code == 200:
        article = Article(article_url)
        article.download()
        article.parse()

        print(f"Title: {article.title}")
        print(f"Text: {article.text}")

    else:
        print(f"Failed to fetch article at {article_url}")
except Exception as e:
    print(f"Error occurred while fetching article at {article_url}: {e}")
```

Sample output:

Title: Meta claims its new AI supercomputer will set records
Text: Ryan is a senior editor at TechForge Media with over a decade of experience covering the latest technology and interviewing leading industry figures. He can often be sighted at tech conferences with a strong coffee in one hand and a laptop in the other. If it's geeky, he's probably into it. Find him on Twitter (@Gadget_Ry) or Mastodon (@gadgetry@techhub.social)

Meta (formerly Facebook) has unveiled an AI supercomputer that it claims will be the world's fastest.

Hugging Face

While the term "Hugging Face" might evoke images of a friendly emoji 🤗 for many, within the technological community, it represents something far more significant: a central hub akin to the "GitHub" for machine learning (ML), dedicated to the collaborative development, training, and deployment of natural language processing (NLP) and ML models through open source collaboration.

The standout feature of Hugging Face lies in its provision of **pre-trained models**. This key innovation means that developers no longer need to initiate their projects from the ground up; instead, they can leverage these ready-made models, adjusting them to fit their specific requirements, thereby streamlining the development workflow.

Hugging Face serves as a vital gathering place for data scientists, researchers, and ML engineers to share insights, solicit support, and contribute to the broader open source movement. Identifying itself as "*the AI community for building the future*," Hugging Face's ethos is deeply rooted in community-driven advancement.

The platform's rapid expansion can also be attributed to its user-friendly design, which welcomes both beginners and seasoned professionals alike. By striving to amass the most extensive collection of NLP and ML resources, Hugging Face is on a mission to democratize AI technology, making it widely available to an international audience.

History of Hugging Face

Hugging Face began its journey in 2016 as a joint American-French venture, initially focusing on creating an AI-driven chatbot designed to engage with teenagers. The turning point for the company came when it decided to share the underlying model of its chatbot with the world, a move that shifted its trajectory toward providing the AI field with robust, easily accessible tools.

The release of the transformative Transformers library in 2018 stands as a milestone in Hugging Face's history, introducing the AI community to pre-trained models such as BERT and GPT, which quickly became foundational tools for NLP tasks.

In the years that followed, Hugging Face has profoundly reshaped the machine learning landscape. Its commitment to open source collaboration has spurred a wave of innovation in NLP, fostering a culture of shared growth and technological advancement.

Hugging Face has evolved into a central hub for the exchange of models and datasets, accelerating both research and practical advancements in AI.

Key Components of Hugging Face

Hugging Face has emerged as a foundational pillar in the field of natural language processing (NLP), offering a suite of tools and resources that cater to diverse linguistic processing requirements. Here's an overview of its core components and functionalities.

Transformers Library

At the heart of Hugging Face is the Transformers library, a collection of cutting-edge machine learning models tailored for NLP tasks. This library includes a wide range of pre-trained models designed for text analysis, content generation, language translation, and summary creation, among other applications. The introduction of the "`pipeline()`" method simplifies the application of these complex models to practical scenarios, offering an intuitive API for a variety of NLP tasks. This library is pivotal for democratizing access to advanced NLP technologies, allowing users to easily customize and deploy sophisticated models.

Hugging Face Hub

The Hugging Face Hub serves as a dynamic online repository, boasting an impressive collection of more than **350,000 models, 75,000 datasets, and 150,000 demonstration applications** (known as Spaces) at the time of writing the book, all of which are open source and freely accessible. This platform is designed as a collaborative ecosystem, enabling individuals to discover, experiment, work together, and develop machine learning technologies. Acting as a pivotal gathering spot, the Hub facilitates the exploration and creation of machine learning projects, encouraging open collaboration and innovation within the community.

The Hugging Face Hub incorporates Git-based repositories for version-controlled management of all related files. This includes the following:

- **Models:** A comprehensive collection of cutting-edge models tailored for NLP, vision, and audio processing tasks

- **Datasets:** An extensive assortment of data across various fields and types, supporting diverse machine learning projects

- **Spaces:** Interactive applications that allow for the direct demonstration of machine learning models within a web browser

Additionally, the Hub is equipped with features such as versioning, commit history, diffs, branches, and integration with over a dozen libraries. For a deeper understanding of these functionalities, the Repositories documentation provides detailed insights.

CHAPTER 7   HARNESSING PYTHON 3.11 AND PYTHON LIBRARIES FOR LLM DEVELOPMENT

Model Hub

Serving as the community's hub, the Model Hub (Figure 7-1) is where users can explore and share a plethora of models and datasets. This feature promotes a collaborative environment for NLP development, enabling practitioners to contribute their own models and benefit from the collective wisdom of the community. The Model Hub is easily navigable on the Hugging Face website, featuring various filters to help users find models suited to specific tasks. This hub is instrumental in fostering a dynamic, evolving ecosystem where new models are regularly added and refined.

Figure 7-1. Model Hub - Hugging Face available models

Within the extensive library of over 200,000 models, you have access to a broad spectrum of functionalities, including the following:

- **Natural Language Processing (NLP):** This encompasses a variety of tasks such as language translation, content summarization, and the creation of text. These capabilities form the essence of what platforms like OpenAI's GPT-3 provide through ChatGPT.

- **Audio Processing:** Here, you can engage in operations like automatic speech recognition, detecting when someone is speaking (voice activity detection), or converting text into speech.

- **Computer Vision:** These models enable computers to interpret and understand visual information from the world. Applications include estimating the distance to objects (depth estimation), categorizing images, and transforming one image to another. Such technology is crucial for developments like autonomous vehicles.

- **Multimodal Models:** These advanced models are capable of processing and generating outputs from multiple data types – text, images, and audio. This versatility allows for a wide range of applications across different media.

Tokenizers

Essential for the preprocessing of text, tokenizers break down language into manageable pieces, or tokens, which are then used by machine learning models to understand and generate human language. These tokens can represent **words, subwords, or characters** and are crucial for converting text into a machine-readable format. Hugging Face's tokenizers are optimized for compatibility with the Transformers library, ensuring efficient text preprocessing for a variety of languages and text formats.

Datasets Library

The Hub features a diverse collection of over 5,000 datasets spanning more than 100 languages, suitable for a wide array of applications in NLP, computer vision, and audio analysis. It streamlines the process of discovering, downloading, and contributing datasets. To enhance user experience, each dataset is presented with comprehensive documentation through Dataset Cards and an interactive Dataset Preview, allowing for in-browser exploration.

The datasets library facilitates a programmatic approach to interacting with these datasets, making it straightforward to integrate them into your projects. This library supports efficient data handling, enabling access to even the largest datasets that exceed your local storage capacity through streaming technology.

CHAPTER 7 HARNESSING PYTHON 3.11 AND PYTHON LIBRARIES FOR LLM DEVELOPMENT

Example App:

First, sign up for an account on Hugging Face and then install the required packages (note that these packages are required by the selected model and could be different for another model).

The code begins by installing necessary Python packages using "pip". These packages include "torch" (PyTorch), "huggingface_hub", "torch accelerate", "torchaudio", "datasets", "transformers", and "pillow" (PIL – Python Imaging Library). These packages are essential for working with deep learning models, datasets, and image processing.

```
!pip install torch
!pip install --upgrade huggingface_hub
!pip install torch accelerate torchaudio datasets
!pip install --upgrade transformers
!pip install git+https://github.com/huggingface/transformers.git
!pip install pillow
```

After installing the required packages, the code imports necessary modules and functions from these packages. Key imports include "huggingface_hub" for interacting with the Hugging Face model hub, "transformers" for accessing pre-trained models, "PIL.Image" for handling images, "requests" for making HTTP requests to fetch images, "torch.nn" for neural network operations, and "matplotlib.pyplot" for plotting images.

```
from transformers import SegformerImageProcessor, AutoModelForSemanticSegmentation
from PIL import Image
import requests
import matplotlib.pyplot as plt
import torch.nn as nn
```

Log in to Hugging Face Model Hub: The code logs into the Hugging Face model hub using the "login()" function. This step is necessary if you plan to use private models or datasets hosted on the Hugging Face platform.

```
from huggingface_hub import login
login()
```

CHAPTER 7   HARNESSING PYTHON 3.11 AND PYTHON LIBRARIES FOR LLM DEVELOPMENT

What happens next in the code of the app:

- **Initialize Image Processor and Model:** It initializes an image processor ("SegformerImageProcessor") and a semantic segmentation model ("AutoModelForSemanticSegmentation"). The models are loaded from the Hugging Face model hub using the specified model names ("mattmdjaga/segformer_b2_clothes"). These models are pre-trained on large datasets and can be fine-tuned or used for inference tasks like semantic segmentation.

- **Fetch Image:** The code retrieves an image from a specified URL using the "requests" module and opens it using the "Image.open()" function from PIL.

- **Preprocess Image:** It preprocesses the fetched image using the initialized image processor ("processor"). This step prepares the image for input to the semantic segmentation model.

- **Model Inference:** The preprocessed image is fed as input to the semantic segmentation model ("model") to obtain segmentation predictions. The "model()" function returns logits, which represent the raw predictions of the model before applying any activation function.

- **Upsample Logits:** The obtained logits are upsampled to the original image size using bilinear interpolation. This ensures that the segmentation predictions match the dimensions of the original image.

- **Extract Segmentation Mask:** The code extracts the predicted segmentation mask from the upsampled logits by applying an argmax operation along the channel dimension. This identifies the class with the highest probability for each pixel, effectively producing a segmentation mask.

- **Display Segmentation Mask:** Finally, the predicted segmentation mask is displayed using "matplotlib.pyplot.imshow()". This visualizes the segmentation mask, highlighting regions corresponding to clothes in the original image.

```python
processor = SegformerImageProcessor.from_pretrained("mattmdjaga/segformer_b2_clothes")
model = AutoModelForSemanticSegmentation.from_pretrained("mattmdjaga/segformer_b2_clothes")

url = "https://www.telegraph.co.uk/content/dam/luxury/2018/09/28/L1010137_trans_NvBQzQNjv4BqZgEkZX3M936N5BQK4Va8RWtTOgK_6EfZT336f62EI5U.JPG"

image = Image.open(requests.get(url, stream=True).raw)
inputs = processor(images=image, return_tensors="pt")

outputs = model(**inputs)
logits = outputs.logits.cpu()

upsampled_logits = nn.functional.interpolate(
    logits,
    size=image.size[::-1],
    mode="bilinear",
    align_corners=False,
)

pred_seg = upsampled_logits.argmax(dim=1)[0]
plt.imshow(pred_seg)
```

In summary, the code segment demonstrates how to use a pre-trained semantic segmentation model to segment clothing in an image fetched from a URL, and visualize the segmentation results.

OpenAI API

The OpenAI API acts as a gateway to the advanced machine learning capabilities developed by OpenAI, enabling seamless integration of state-of-the-art AI functionalities into your applications. Essentially, this API functions as a conduit, granting access to OpenAI's sophisticated algorithms, which include capabilities for text understanding, generation, and even code creation, all without requiring deep technical knowledge of the models that power these abilities.

Features of the OpenAI API

This section provides an in-depth look at the key features that make the OpenAI API a versatile and powerful tool for a wide range of applications. Additionally, we will discuss its user-friendly integration options, customizable settings, and scalability, which allow for seamless implementation across various platforms and projects.

Pre-trained Models

The API offers access to a variety of pre-trained models, which have been developed and refined by OpenAI. These models, including versions of GPT-4, GPT-3.5, and others, are ready to be deployed for tasks ranging from text and code generation to image creation and audio transcription. This collection also includes specialized models for embeddings, content moderation, and more, all trained on vast datasets with substantial computational resources, making sophisticated machine learning accessible to a wider audience.

Key models include the following:

- **GPT-4 and GPT-3.5:** Advanced models capable of generating and understanding both text and code
- **DALL-E:** Generates and alters images based on textual descriptions
- **Whisper:** Converts spoken language into text
- **Embeddings and Moderation Models:** For transforming text into numerical representations and identifying sensitive content, respectively

Customization Through Fine-Tuning

The API allows for the customization of pre-trained models via fine-tuning, adapting them to better fit specific requirements. This process involves retraining models on your own data, enhancing their performance on particular tasks, which can lead to cost savings and improved efficiency for targeted applications.

User-Friendly API Interface

Designed with simplicity in mind, the OpenAI API is accessible even to those new to data science, featuring straightforward documentation and examples that guide users through incorporating AI into their projects effortlessly.

Scalable Infrastructure

The infrastructure behind the OpenAI API, including robust Kubernetes clusters, ensures scalability to accommodate projects of any size. This scalability is crucial for supporting the deployment of large models and accommodating the growth of user projects over time.

Industry Applications of the OpenAI API

The versatility of the OpenAI API is reflected in its wide range of applications across various industries:

- **Chatbots and Virtual Assistants:** Leveraging models like GPT-4, the API can power conversational agents that offer realistic and engaging user interactions, suitable for customer service and interactive applications.

- **Sentiment Analysis:** By analyzing textual data, such as customer feedback or social media posts, the API can provide valuable insights into public sentiment and customer satisfaction.

- **Image Recognition:** With models like CLIP, the API extends its capabilities to visual tasks, enabling object detection and image classification, which have applications in fields from retail to healthcare.

- **Text Comparison:** The OpenAI API provides a text comparison feature that leverages the text-embedding-ada-002 model, a sophisticated second-generation embedding model. This model assesses the similarity or dissimilarity between pieces of text by mapping them into vector spaces and measuring the distances between these vectors. **The greater the distance, the less similar the texts are considered to be.** This embedding functionality supports a variety of applications including text clustering, identifying differences and relevancies between texts, generating recommendations, analyzing sentiments, and performing text classification, with costs based on the amount of text processed. While OpenAI's documentation acknowledges the availability of first-generation embedding models, it highlights the superior efficiency and cost-effectiveness of the latest model. However, it also cautions users about the potential for the model to exhibit social biases toward certain groups, as identified in various tests.

- **Code Completion:** Regarding code development, the OpenAI API includes a code completion service powered by OpenAI Codex, an advanced model trained on a vast corpus of natural language and billions of lines of code from public sources. Currently in a limited beta phase and offered free of charge at the time of writing, this service supports a wide range of programming languages. The service relies on models like code-davinci-002 and code-cushman-001 to facilitate the insertion of code lines or the generation of code blocks in response to prompts provided by the user. While the code-cushman-001 model is noted for its speed, the code-davinci-002 model is distinguished by its comprehensive capabilities, particularly in the area of code auto-completion.

- **Gaming and Reinforcement Learning:** The API supports the development of AI that can navigate gaming environments, either by playing autonomously or aiding human players, showcasing its potential in reinforcement learning applications.

- **Image Generation:** This is one of the most intuitive features of the OpenAI API. Based on the DALL-E image model, the OpenAI API's image functionality features endpoints for generating, editing, and creating image variations from natural language prompts. Although it doesn't yet have advanced features, its outputs are more impressive than those of generative art models like Midjourney and Stable Diffusion. While hitting the image generation endpoint, you only need to supply a prompt, image size, and image count. But the image editing endpoint requires you to include the image you wish to edit and an RGBA mask marking the edit point in addition to the other parameters. The variation endpoint, on the other hand, only requires the target image, the variation count, and the output size. The endpoint for generating images enables the creation of unique visuals from a textual description. With DALL-E 3, these images can be produced in dimensions of 1024x1024, 1024x1792, or 1792x1024 pixels.

The OpenAI API stands as a powerful tool for integrating advanced AI into a multitude of projects, democratizing access to machine learning innovations and fostering the development of intelligent, responsive, and personalized technologies.

Simple Example of a Connection to the OpenAI API

The application utilizes the OpenAI platform to generate a summarized biography of a specified individual. Users input the name of the person they are interested in, and the application sends a request to OpenAI's chat API. The API uses the GPT-4 model to understand the user's request and generates a concise summary of the person's biography. The summarized biography is then displayed to the user, providing them with a quick overview of the individual's life and achievements.

Example App:

```
pip install openai
from openai import OpenAI

client = OpenAI(
  api_key='YOUR API KEY'  # this is also the default, it can be omitted
)

person = input() # Write the name of the person you are interested in.
response = client.chat.completions.create(
  model="gpt-4",
  messages=[
    {"role": "system", "content": "You are a person biography summarizer."},
    {"role": "user", "content": f"Summarize this biography for me {person}"},
  ]
)
print(response.choices[0].message.content)
```

CHAPTER 7 HARNESSING PYTHON 3.11 AND PYTHON LIBRARIES FOR LLM DEVELOPMENT

Instructions

1. **Install the OpenAI package using pip:**

   ```
   pip install openai==1.35.7
   ```

2. **Import the OpenAI module:**

   ```
   from openai import OpenAI
   ```

3. **Set up the OpenAI client with your API key:**

   ```
   client = OpenAI(
     api_key='YOUR API KEY'  # this is also the default, it can
     be omitted
   )
   ```

4. **Prompt the user to input the name of the person they are interested in:**

   ```
   person = input() # Write the name of the person you are
   interested in.
   ```

5. **Send a request to OpenAI's chat API to summarize the biography of the specified person:**

   ```
   response = client.chat.completions.create(
     model="gpt-4",
     messages=[
       {"role": "system", "content": "You are a person biography
       summarizer."},
       {"role": "user", "content": f"Summarize this biography for me
       {person}"},
     ]
   )
   ```

   After setting up the OpenAI client with the API key, the Python script sends a request to OpenAI's chat API to generate a summary of the biography for the specified person.

Here's a breakdown of the request:

- **Model Selection:** The model parameter is set to "gpt-4", indicating the version of the GPT (Generative Pre-trained Transformer) model to be used for generating the summary. In this case, it's specified as "gpt-4".

- **Messages:** The messages parameter is a list containing two dictionaries, each representing a message in the conversation between the user and the system.

- **The first dictionary represents a message from the system** with the role set as "system" and the content as "You are a person biography summarizer." This message informs the AI model about the context of the conversation.

- **The second dictionary represents a message from the user** with the role set as "user" and the content dynamically generated using f-string formatting. It prompts the AI to summarize the biography of the specified person. The name of the person is included in the content of the message.

Once the request is sent to the OpenAI API, the response contains the generated summary, which is then extracted and printed using the print statement.

6. **Print the summarized biography:**

```
print(response.choices[0].message.content)
```

Cohere

Cohere, established in 2019 and headquartered across Toronto and San Francisco with additional offices in Palo Alto and London, operates as a global technology enterprise with a focus on artificial intelligence solutions for businesses, particularly through the development of sophisticated large language models. The company's inception was the collective effort of founders **Aidan Gomez, Ivan Zhang, and Nick Frosst**, all of whom share an academic background from the University of Toronto.

The roots of Cohere trace back to a transformative moment in AI research when, in 2017, Gomez, as part of the Google Brain team, co-authored the groundbreaking paper "Attention is All You Need," introducing the transformer model that revolutionized natural language processing tasks. This pioneering work laid the groundwork for Cohere's establishment by Gomez, Frosst, and Zhang, who previously collaborated at FOR.ai.

Since its launch, Gomez has led Cohere as CEO, with the company marking a significant milestone by appointing Martin Kon, the former CFO of YouTube, as president and COO in late 2022. Cohere's technological advancements have garnered significant support, including a strategic partnership with Google Cloud in 2021 to leverage Google's infrastructure and TPUs for product development.

Cohere's commitment to innovation is further evident in the June 2022 establishment of Cohere For AI, a nonprofit research initiative aimed at advancing open source machine learning research under the leadership of Sara Hooker, formerly of **Google Brain**. The company also made strides in language processing with a multilingual model capable of understanding over 100 languages, a significant achievement in making non-English language processing more accessible.

Collaborations with Oracle, McKinsey, and LivePerson throughout 2023 have expanded Cohere's footprint in offering generative AI services and customized language models to businesses, enhancing automation and operational efficiency across various industries. Furthermore, Cohere's proactive engagement with ethical AI practices is demonstrated by its adherence to voluntary measures laid out by the White House and Canada's code of conduct for AI development and management.

Positioned as a contender to OpenAI, Cohere's technology suite serves enterprises seeking to implement AI for a range of applications, from chatbots and search engines to content moderation and data analysis. The platform's API facilitates integration with major cloud services, ensuring versatility and broad application.

Cohere Models

Cohere offers a diverse array of models tailored to meet a broad spectrum of needs. For those seeking a more bespoke solution, there is the option to custom-train a model to align precisely with particular requirements.

Command

The Command model serves as Cohere's primary generation tool, designed to interpret and execute textual commands or prompts from users. This model is not only adept at generating text in response to instructions but also possesses conversational abilities, making it ideal for powering chat-based applications.

Embed

The Embed models provide functionality for generating text embeddings or for classifying text according to a set of criteria. These embeddings are useful for a range of tasks, such as measuring the semantic similarity between sentences, selecting the sentence most likely to succeed another, or sorting user feedback into categories.

Additionally, the Classify function within the Embed models supports various classification or analytical tasks. The Representation model enhances these capabilities with additional support functions, including language detection for inputs.

Rerank

Lastly, the Rerank model is designed to refine and optimize the outputs of existing models by reordering their results based on specific criteria. This functionality is particularly beneficial for enhancing the efficacy of search algorithms.

Example App for Sentiment Analysis

Sign up on the Cohere website and get your API key, then install Cohere SDK in your notebook or terminal with the following command: `pip install cohere`, then use the following code:

```
import cohere
from cohere.responses.classify import Example
co = cohere.Client('GHviIR5p9NC7kNzRf383ykOxU2Y9LQVbSAvAdNSj')
examples=[
  Example("Dermatologists don't like her!", "Spam"),
  Example("'Hello, open to this?'", "Spam"),
  Example("I need help please wire me $1000 right now", "Spam"),
  Example("Nice to know you ;)", "Spam"),
  Example("Please help me?", "Spam"),
```

CHAPTER 7   HARNESSING PYTHON 3.11 AND PYTHON LIBRARIES FOR LLM DEVELOPMENT

```
  Example("Your parcel will be delivered today", "Not spam"),
  Example("Review changes to our Terms and Conditions", "Not spam"),
  Example("Weekly sync notes", "Not spam"),
  Example("'Re: Follow up from today's meeting'", "Not spam"),
  Example("Pre-read for tomorrow", "Not spam"),
]
user_input = input()
inputs=[
  user_input
]
response = co.classify(
  model = 'large',
  inputs=inputs,
  examples=examples,
)
print(response.classifications)
```

Output:

```
Looking forward to your email
[Classification<prediction: "Not spam", confidence: 0.5914174, labels:
{'Not spam': LabelPrediction(confidence=0.5914174), 'Spam': LabelPrediction
(confidence=0.40858263)}>]
```

This Python code uses the "cohere" library to perform sentiment analysis. Here's a breakdown of what the code does:

1. **Import necessary modules:**

    - "cohere": The main library for interfacing with the Cohere API.

    - "Example" from "cohere.responses.classify": This is used to define example inputs with their corresponding labels for training the model.

2. **Create a "cohere.Client" object:**

    - A client object is created to interact with the Cohere API. It requires an API key for authentication.

CHAPTER 7   HARNESSING PYTHON 3.11 AND PYTHON LIBRARIES FOR LLM DEVELOPMENT

3. **Define a list of example inputs with their corresponding labels:**

    - Each example is an instance of the "Example" class from the "cohere.responses.classify" module.

    - These examples are used for training the classification model. They are labeled as either *"Spam" or "Not spam"*.

4. **Prompt the user for input:**

    - The "input()" function prompts the user to enter some text, which will be classified as either "Spam" or "Not spam".

5. **Create a list containing the user's input:**

    - The user's input is stored in a list called "inputs".

6. **Use the "co.classify()" method to classify the user's input:**

    - The "co.classify()" method takes several parameters:

    - "model": Specifies the model to be used for classification. Here, it's set to "large".

    - "inputs": The list of inputs to be classified. In this case, it contains only the user's input.

    - "examples": The list of example inputs and their labels used to train the model.

    - This method returns a response object.

7. **Print the classifications:**

    - The "response.classifications" attribute contains the classifications for the input text.

    - This will print out the classification result, indicating whether the input is categorized as "Spam" or "Not spam".

Pinecone

In today's digital era, where rapid access to and storage of diverse information forms are paramount, traditional relational databases fall short in managing varied data types like documents, key-value pairs, and graphs. Enter the era of vector databases, a cutting-edge solution that employs vectorization for enhanced search capabilities, efficient storage, and in-depth data analysis.

Among these innovative databases, Pinecone stands out as a leading vector database widely recognized for its ability to tackle issues related to complexity and dimensionality. Pinecone, a vector database engineered for the cloud, excels in managing high-dimensional vector data. At its core, Pinecone leverages the Approximate Nearest Neighbor (ANN) search algorithm to swiftly find and rank the closest matches within vast datasets.

This guide will delve into the intricacies of Pinecone, highlighting its key features, challenges it addresses, and practical applications.

How Vector Databases Operate

Unlike traditional databases that seek exact matches to queries, vector databases aim to find the query's closest vector match. Utilizing ANN search, these databases deliver near-accurate results swiftly and efficiently.

Pinecone operates as a vector database hosted in the cloud, tailored for use with machine learning projects. If the term "vector database" seems complex, let's break it down further.

Here's a closer look at the operational mechanics of vector databases:

1. **Data Conversion and Indexing:** Initially, data is transformed into vectors, with indexes created for quick searches.

2. **Search and Comparison:** The database compares the query vector against indexed vectors using similarity measures, identifying the closest matches.

3. **Result Refinement:** The most relevant results are then refined and presented, based on the similarity assessments.

What Exactly Is a Vector Database?

Vector databases are specialized storage systems crafted to manage vector embeddings – these are compact arrays of numbers that encapsulate textual information. In machine learning contexts, these embeddings are crucial for grasping the nuances of word meanings and their interrelations. By indexing these vectors, the database can swiftly locate and retrieve items based on their similarity, rendering it a powerful tool for applications in natural language processing and AI technologies.

Pinecone's Features

Pinecone distinguishes itself with a suite of features that cater to the needs of modern data infrastructure:

- **Rapid and Current Vector Search:** Pinecone ensures minimal query latency, capable of handling billions of items efficiently. Its indexes are continuously updated, ensuring access to the latest data.
- **Filtered Searches:** It allows for the integration of vector searches with metadata filters, yielding more precise and quicker outcomes.
- **Dynamic Data Updates:** Unlike traditional vector indexes requiring comprehensive re-indexing, Pinecone supports instant data modifications.
- **Data Management:** Routine data backups and the option to save specific indexes as "collections" for future use are part of Pinecone's offering.
- **Accessible API:** The user-friendly API facilitates the development of vector search applications, compatible with various programming languages.
- **Cost-Effective:** Thanks to its cloud-native architecture, Pinecone is not only efficient but also offers a pay-per-use pricing model.

Despite its advantages, Pinecone faces challenges such as the following:

- Integration complexities with other systems over time.
- Ensuring data privacy through robust security mechanisms.

- The interpretability of vector-based models remains a challenge, making it difficult to understand the rationale behind certain data relationships.

Practical Applications

Pinecone's utility spans across numerous sectors:

- **Audio/Text Searches:** Offers advanced search capabilities for text and audio data

- **Natural Language Processing:** Facilitates NLP applications including document classification and sentiment analysis

- **Recommendation Systems:** Enhances user experiences through personalized recommendation engines

- **Image and Video Retrieval:** Speeds up the process of identifying relevant image and video content

- **Time Series Analysis:** Assists in recognizing patterns within time-series data, aiding in recommendations and data clustering

Pinecone represents a significant leap forward in database technology, offering a scalable, efficient, and versatile vector database solution. With its high-performance search functionalities and ability to handle complex vector data, Pinecone stands as a formidable tool for businesses seeking to navigate the challenges of modern data management.

Example App:

First, sign up on the Pinecone website and get your API key, then install the Pinecone Client with the following command: `pip install pinecone-client`. Overall, the following code snippet demonstrates the basic usage of Pinecone for creating an index, adding vectors to it, and querying for similar vectors based on specific criteria.

Use the following code:

```
from pinecone import Pinecone

pc = Pinecone(api_key="YOUR-API-KEY")
index = pc.Index("test-index")
```

```python
index.upsert(
    vectors=[
        {
            "id": "vec1",
            "values": [0.1, 0.1, 0.1, 0.1, 0.1, 0.1, 0.1, 0.1],
            "metadata": {"genre": "drama"}
        }, {
            "id": "vec2",
            "values": [0.2, 0.2, 0.2, 0.2, 0.2, 0.2, 0.2, 0.2],
            "metadata": {"genre": "action"}
        }, {
            "id": "vec3",
            "values": [0.3, 0.3, 0.3, 0.3, 0.3, 0.3, 0.3, 0.3],
            "metadata": {"genre": "drama"}
        }, {
            "id": "vec4",
            "values": [0.4, 0.4, 0.4, 0.4, 0.4, 0.4, 0.4, 0.4],
            "metadata": {"genre": "action"}
        }
    ],
    namespace= "ns1"
)
index.query(
    namespace="ns1",
    vector=[0.3, 0.3, 0.3, 0.3, 0.3, 0.3, 0.3, 0.3],
    top_k=2,
    include_values=True,
    include_metadata=True,
    filter={"genre": {"$eq": "action"}}
)
```

Output:

```
{'matches': [{'id': 'vec2',
              'metadata': {'genre': 'action'},
              'score': 1.0,
```

```
              'values': [0.2, 0.2, 0.2, 0.2, 0.2, 0.2, 0.2, 0.2]},
             {'id': 'vec4',
              'metadata': {'genre': 'action'},
              'score': 1.0,
              'values': [0.4, 0.4, 0.4, 0.4, 0.4, 0.4, 0.4, 0.4]}],
'namespace': 'ns1',
'usage': {'read_units': 6}}
```

Here's what happens step by step in the app:

- **Importing:** After installing the package, it imports the "Pinecone" class from the "pinecone" module.

- **Initialization:** An instance of the "Pinecone" class is created with an API key. This API key is used for authentication and authorization to access the Pinecone service.

- **Creating an Index:** A new index named "test-index" is created using the "pc.Index()" method. An index in Pinecone is a storage structure that organizes and allows efficient querying of vectors.

- **Upserting Vectors:** Vectors are added to the index using the "upsert()" method. Each vector consists of an ID, a list of values representing the vector itself, and optional metadata. In this example, four vectors are added, each with an ID, a list of values, and metadata specifying the genre.

- **Querying the Index:** The `query()` method is used to search for vectors similar to a given query vector. In this case, a query vector "[0.3, 0.3, 0.3, 0.3, 0.3, 0.3, 0.3, 0.3]" is provided. The "top_k" parameter specifies that the top two most similar vectors should be returned. "include_values" and "include_metadata" parameters are set to "True" to include the vector values and metadata in the query result, respectively. Additionally, a filter is applied to retrieve only vectors with the "action" genre metadata.

Lamini.ai

Lamini is at the forefront of artificial intelligence innovation, introducing an advanced AI-powered large language model (LLM) platform designed to transform enterprise software development. Leveraging the power of generative AI and machine learning, Lamini offers a dynamic tool that automates workflows, enriches the software development lifecycle, and elevates productivity levels. What sets this platform apart is its capacity to equip developers with sophisticated tools and features, enabling the crafting of private, tailored models that surpass the efficiency, speed, and usability of conventional LLMs.

Lamini's Operational Mechanics

Lamini integrates the cutting-edge functionalities of GPT-3 and ChatGPT into the domain of enterprise software development, granting developers access to features that enhance speed, incorporate enterprise-specific functionalities, and ensure the secure handling of private data. This platform customizes models to align with each organization's unique data and operational requirements, improving the AI-generated outputs' pertinence and effectiveness.

Key aspects of Lamini's operation include the following:

- Selection of an initial base model to serve as a foundation
- Retrieval-augmented training to expand the model's knowledge
- Efficient fine-tuning to achieve targeted customization
- Advanced data generation tools
- Automatic performance evaluation mechanisms

Lamini's Features, Functionalities, and Advantages

Lamini boasts an array of features and functionalities that offer substantial benefits to businesses, including the following:

- **Simplified Model Training:** With Lamini's toolkit, training new LLMs requires minimal coding effort, facilitating smoother deployment.

- **Hands-Off Infrastructure Management:** Lamini oversees hosting and computing, allowing developers to focus on innovation.

- **Guaranteed Data Privacy:** Organizations can operate within their secure environments, maximizing data usage while maintaining privacy.

- **State-of-the-Art Model Foundations:** Lamini builds on the latest models from industry leaders like Hugging Face and OpenAI.

- **Tailored Customization:** Models are customized to fit the linguistic nuances and specific requirements of each business.

- **Deployment Versatility:** Models can either be hosted on Lamini's cloud or deployed privately, on-premises, or in a private cloud.

Applications and Use Cases for Lamini

Lamini is versatile, supporting a broad spectrum of applications and use cases, enabling businesses to

- **Deploy AI-driven chatbots for customer service** that understand and operate within the context of company-specific terminologies and processes

- **Produce and refine content with AI tools** that align with brand guidelines and voice

- **Simplify the coding and debugging workflow** through AI models that are adept at navigating a company's unique code base and programming methodologies

Example App:

In order to use Lamini, first you need to install and upgrade it in your notebook or terminal and to sign up on their website in order to get an API key. Install and upgrade Lamini by using the following commands:

```
pip install lamini
pip install --upgrade lamini
```

Then use the following code:

```python
import lamini
lamini.api_key = "YOUR-API-KEY"

from lamini import LaminiClassifier

llm = LaminiClassifier()

prompts={
  "cat": "Cats are generally more independent and aloof than dogs, who are
  often more social and affectionate.",
  "dog": "Dogs are more pack-oriented and tend to be more loyal to their
  human family.",
}

llm.add_data_to_class("dog", ["woof", "group oriented"])
llm.add_data_to_class("cat", ["Oh, I prefer to do stuff on my own than
dogs", "meow"])   # list of examples is valid too

llm.prompt_train(prompts)

llm.predict(["I'm more independent than dogs", "woof"])
```

Sample output:

```
0%|          | 0/10 [00:34<?, ?it/s]
ERROR:lamini.classify.llama_classifier:Failed to generate examples for
class cat
ERROR:lamini.classify.llama_classifier:string indices must be integers
  0%|          | 0/10 [00:00<?, ?it/s]ERROR:lamini.classify.llama_
classifier:string indices must be integers
ERROR:lamini.classify.llama_classifier:Consider rerunning the generation
task if the error is transient, e.g. 500
  0%|          | 0/10 [00:33<?, ?it/s]
ERROR:lamini.classify.llama_classifier:Failed to generate examples for
class dog
ERROR:lamini.classify.llama_classifier:string indices must be integers
ERROR:lamini.classify.llama_classifier:string indices must be integers
```

```
ERROR:lamini.classify.llama_classifier:Consider rerunning the generation
task if the error is transient, e.g. 500
100%|████████████████████████| 2/2 [00:01<00:00,  1.62it/s]
```

['cat', 'dog']

Explanation

The preceding code utilizes the Lamini library for text classification, specifically for classifying text into different categories or classes. Here's a detailed breakdown of what each part of the code does:

1. **Installation of Lamini**

   The first two lines (`"!pip install lamini"` and `"!pip install --upgrade lamini"`) are using "pip", a package management system for Python, to install the Lamini library and ensure it's up to date.

2. **Importing LaminiClassifier**

   "`from lamini import LaminiClassifier`" imports the "LaminiClassifier" class from the "lamini" module, which is a part of the Lamini library.

3. **Creating an Instance of LaminiClassifier**

   "`llm = LaminiClassifier()`" creates an instance of the "LaminiClassifier" class, which will be used for text classification tasks.

4. **Defining Prompts**

   - "`prompts`" is a dictionary containing example texts associated with different categories. Each key-value pair in the dictionary represents a category label and its corresponding example text.

   - In this case, there are two categories: "cat" and "dog", with example texts describing characteristics or behaviors associated with each category.

5. **Adding Data to Classes**
    - The "llm.add_data_to_class()" method is used to add training data to each class. It associates the provided text examples with the respective class label.

        For example, "llm.add_data_to_class("dog", ["woof", "group oriented"])" adds the provided texts "woof" and "group oriented" to the "dog" class.

    - Similarly, "llm.add_data_to_class("cat", ["Oh, I prefer to do stuff on my own than dogs", "meow"])" adds the provided texts to the "cat" class.

6. **Training the Model with Prompts**

    "llm.prompt_train(prompts)" trains the Lamini classifier using the provided prompts and associated example texts. This step involves training the model to recognize patterns in the text data and learn to classify new text inputs into the appropriate categories based on the provided examples.

7. **Making Predictions**
    - "llm.predict(["I'm more independent than dogs", "woof"])" makes predictions on the provided text inputs.
    - In this case, the model predicts the category labels for the given texts "I'm more independent than dogs" and "woof", based on the training it received using the prompts and example texts.

Note Environmental variables

It's recommended to add your API key as an environmental variable for security purposes.

Data Collection, Cleaning, and Preparation of Python Libraries

The effectiveness of large language models (LLMs) can be attributed significantly to their vast size, as they are developed using extensive datasets. This extensive training allows them to grasp a wide array of themes, genres, and languages more comprehensively than smaller models limited by less diverse data.

The principle guiding this approach is simple: more data equates to better performance. Datasets like C4, The Pile, The Bigscience Roots Corpus, and OpenWebText have played pivotal roles in expanding the volume of training data by aggregating and refining vast text collections from web crawls, aimed at enhancing LLM pre-training.

However, the sheer cost of manually reviewing and refining such large datasets means many contain quality issues. The significance of this extends beyond just technical performance metrics such as perplexity and validation loss; it also means that the models can inadvertently learn and propagate the biases present in their training materials. Consequently, understanding the nature and composition of these datasets is not just a technical necessity but a research challenge of its own.

As data acts as the cornerstone for LLM development, comprehending and cataloging the specifics of training datasets becomes imperative. This is vital for assessing the value of data in the models' predictions and decisions, especially since the suitability of training data can vary greatly depending on the context of the application. The recommended strategy is to document the potentially problematic aspects of datasets rather than attempting to eliminate them entirely.

In the realm of machine learning, there's typically a similarity or congruence between training and testing (evaluation) data. However, LLMs are trained on what is essentially "raw text," leading to unique challenges in creating non-overlapping training, validation, and test splits, especially in relation to benchmark datasets.

With this understanding, let's explore the management of the colossal datasets necessary for training these advanced models!

Gathering and Preparing Data for Large Language Models

The foundation of large language models (LLMs) lies in the quality and preparation of their pre-training data. Unlike smaller counterparts, the effectiveness and the capabilities of LLMs are largely determined by the richness of their pre-training corpus and the meticulousness of its preprocessing.

Data Acquisition

The journey to developing a robust LLM begins with the collection of a comprehensive natural language corpus, sourced from a wide array of platforms. The diversity and volume of data collected from these sources directly influence the model's proficiency.

Types of Data Used

- **General Data:** The bulk of pre-training corpora for current LLMs comprises general data, which includes content from web pages, books, and dialogues. This type of data is preferred due to its vast availability, variety, and the ease with which it can be accessed, fostering the model's ability to understand and generate human-like text.

- **Specialized Data:** To equip LLMs with the ability to tackle niche problems, researchers also incorporate specialized datasets into the training mix. These can include datasets rich in multilingual content, scientific literature, and programming code, aiming to refine the model's expertise in specific domains.

What Is Data Preprocessing?

Data cleansing, often referred to as data preprocessing, stands as a pivotal phase within the realms of data analysis and the development of machine learning models. This stage is dedicated to the meticulous examination, modification, and rectification of raw data to elevate its quality, precision, and uniformity.

The essence of data cleansing lies in its ability to identify and amend inaccuracies, noise, missing elements, duplications, inconsistencies, and other flaws within the dataset. This refinement process is imperative to prepare the data for subsequent steps in analysis, modeling, and knowledge discovery. The value of a dataset is significantly enhanced when it is relevant to the specific objectives of the model, exhibits diversity, and maintains a high standard of quality.

Given the prevalence of challenges such as missing values, duplications, and noise, large datasets cannot be employed in their raw form for training sophisticated language models. Such datasets necessitate undergoing rigorous cleansing and labeling procedures to render them apt for use in training large language models.

This transformation is crucial for leveraging the full potential of algorithms, computational resources, and other technological advancements in the development of these models. For instance, the initial dataset for GPT-3 spanned an impressive 45TB; however, only 570GB of data, representing about 1% of the original volume, met the high-quality standards required for inclusion in the training corpus after the cleansing process.

Preparing Datasets for Training

The goal of training language models can differ based on their intended use cases, yet there are several crucial practices to adopt to guarantee the training data for LLMs is both clean and reliable.

These practices encompass

- Managing unwanted data
- Deduplication
- Decontamination
- Toxicity and bias control
- Personally identifiable information control
- Prompt control
- Tokenization and vectorization
- Handling of missing data
- Data augmentation

Managing Unwanted Data

Large datasets, despite their extensive size, often contain a significant amount of non-useful content, including nonsensical text and standard template material, such as HTML code or placeholder text (e.g., Lorem ipsum). When gathering text from the Web for the purpose of language model training, particularly for datasets that include multiple languages, filtering out this "junk" data is critical. Before training a model to predict subsequent tokens based on previous ones, it's essential to purify the dataset of such elements.

Tools and methods like **justext and trafilatura** are effective for eliminating standard web page filler while maintaining a balance between minimizing irrelevant content (precision) and retaining all pertinent content (recall). Additionally, leveraging metadata associated with web content can serve as an effective filter.

A simple example with trafilatura:

```python
import trafilatura

def filter_unwanted_data(url, output_file, unwanted_keywords):
    # Download and extract text content from the web page
    downloaded_data = trafilatura.fetch_url(url)
    extracted_text = trafilatura.extract(downloaded_data)

    # Open output file for writing
    with open(output_file, 'w') as output_f:
        # Filter out unwanted data based on specified keywords
        for line in extracted_text.split('\n'):
            if not any(keyword in line for keyword in unwanted_keywords):
                output_f.write(line + '\n')

if __name__ == "__main__":
    # Define URL of the web page to extract content from
    url = 'https://example.com'

    # Define output file path
    output_file_path = 'filtered_content.txt'

    # Define list of unwanted keywords
    unwanted_keywords = ['unwanted1', 'unwanted2', 'unwanted3']

    # Call filter_unwanted_data function to filter unwanted data
    filter_unwanted_data(url, output_file_path, unwanted_keywords)

    print("Unwanted data filtered successfully!")

import trafilatura

def filter_unwanted_data(url, output_file, unwanted_keywords):
    # Download and extract text content from the web page
    downloaded_data = trafilatura.fetch_url(url)
    extracted_text = trafilatura.extract(downloaded_data)
```

```python
    # Open output file for writing
    with open(output_file, 'w') as output_f:
        # Filter out unwanted data based on specified keywords
        for line in extracted_text.split('\n'):
            if not any(keyword in line for keyword in unwanted_keywords):
                output_f.write(line + '\n')
if __name__ == "__main__":
    # Define URL of the web page to extract content from - feel free to
    change the URL
    url = 'https://blog.hootsuite.com/what-is-discord/'

    # Define output file path
    output_file_path = 'filtered_content.txt'

    # Define list of unwanted keywords, feel free to add yours
    unwanted_keywords = ['platform', 'business', 'marketing']

    # Call filter_unwanted_data function to filter unwanted data
    filter_unwanted_data(url, output_file_path, unwanted_keywords)

    print("Unwanted data filtered successfully!")
```

Content in the file:

```
If you work in social media, you may be wondering, "What is Discord — and
wait, why should I care?"
What is the Discord app?
Servers can be public or private spaces. You can join a big community for
people who share a common interest or start a smaller private server for a
group of friends.
How did Discord get started?
Discord launched in 2015, and its initial growth was largely thanks to
its widespread adoption by gamers. However, it wasn't until the COVID-19
pandemic that it began to attract a broader audience.
The company embraced its newfound audience, changing its motto from "Chat
for Gamers" to "Chat for Communities and Friends" in May 2020 to reflect
its more inclusive direction.
Who uses Discord now?
```

CHAPTER 7 HARNESSING PYTHON 3.11 AND PYTHON LIBRARIES FOR LLM DEVELOPMENT

Source: eMarketer

1. Build community

These lfg channels accomplish two things for Fortnite. First, they build a community around the brand by making it easier for fans to connect. And they make it easier for players to use their product.

In this case, Discord doesn't just help Fortnite players connect outside the game. It improves their experience of the product itself.

2. Use roles to customize your audience's Discord experience

(Discord roles are a defined set of permissions that you can grant to users. They're handy for plenty of reasons, including customizing your community's experience on your server)

Here are a few ways to use roles in your server:

- Flair: Use roles to give users aesthetic perks, like changing the color of their usernames or giving them custom icons.
- Custom alerts: Use "@role" in the chat bar to notify all users with the role. This allows you to send messages to specific segments of your audience.
- Role-based channels: Grant users access to exclusive channels open only to users with certain roles.
- VIP roles: Reward paying subscribers or customers with a VIP role. Combined with role-based channels, you can make subscriber-only channels.
- Identity roles: Discord profiles are pretty bare bones. With roles, users can let each other know what their pronouns are or what country they're from.

A server template provides a Discord server's basic structure. Templates define a server's channels, channel topics, roles, permissions, and default settings.

You can use one of Discord's pre-made templates, one from a third-party site, or create your own.

Can I advertise on Discord?

Save time managing your social media presence with Hootsuite. Publish and schedule posts, find relevant conversions, engage the audience, measure results, and more – all from one dashboard. Try it free today.

In this app:

- The `filter_unwanted_data` function takes three parameters: `url` (URL of the web page to extract content from), `output_file` (path to the output file), and `unwanted_keywords` (a list of keywords that represent unwanted data).

- The function downloads the web page content using Trafilatura, extracts text content from it, and filters out lines containing any of the unwanted keywords.

- The filtered text content is then written to the output file.

- In the `__main__` block, the URL of the web page, the output file path, and the list of unwanted keywords are defined.

- The `filter_unwanted_data` function is called with these parameters to filter unwanted data from the web page content.

- A message is printed to indicate that unwanted data has been filtered successfully.

Handling Document Length

In language modeling, where the objective is to generate text based on preceding tokens, excluding overly brief documents (those with fewer than approximately 100 tokens) from the dataset can diminish distractions, fostering a more coherent modeling of textual dependencies. Given the prevalent use of transformer architectures in contemporary language models, preprocessing to divide extensive documents into consistent, manageable segments is beneficial. The following code snippet, from the datasets library, illustrates how to segment large documents into discrete, non-overlapping sections:

Simple example:

```
def segment_text(examples):
    segmented_texts = []
    for text in examples['text']:
        # Break the text into segments of 70 characters
        segmented_texts += [text[j:j + 70] for j in range(0, len(text), 70)]
    return {'segmented_texts': segmented_texts}
```

Text Produced by Machines

A primary objective in the development of language models is to accurately represent the spectrum of human language. Nevertheless, datasets derived from web crawls often incorporate a significant amount of text produced by machines. This includes output from existing language models, text digitized via Optical Character Recognition (OCR), and content that has been machine-translated.

For example, the C4 corpus significantly incorporates data from patents.google.com, which relies on machine translation to convert patents from global patent offices into English. Furthermore, web-based datasets frequently include OCR text from scanned books and documents. Given the inherent imperfections of OCR technology, the resulting text often deviates from the natural distribution of English, exhibiting predictable errors like misspellings and omissions.

Identifying text generated by machines poses considerable challenges and is an area of ongoing research. Nonetheless, tools like the ctrl-detector offer some capability for detecting such machine-generated content. It's crucial, when preparing a dataset for language modeling, to identify, characterize, and document any machine-generated text it may contain.

Removing Duplicate Content

When collecting text data from the Web, it's common to encounter repeated instances of the same text. For instance, research highlighted in "Deduplicating Training Data Makes Language Models Better" discovered that a particular 50-word text sequence appeared in the C4 dataset 60,000 times. Training language models on datasets from which duplicates have been removed not only speeds up the process but also reduces the risk of the models memorizing specific sequences.

Moreover, recent studies indicate that models trained on data containing duplicates are vulnerable to privacy breaches, wherein attackers can prompt the model to reproduce specific sequences, thereby revealing which data it has memorized. The study "Deduplicating Training Data Mitigates Privacy Risks in Language Models" illustrates that the frequency with which models reproduce specific sequences from their training data increases superlinearly with the sequence's prevalence within that data. For example, a sequence appearing ten times is more likely to be generated by the model at a rate 1000 times higher than one that appears only once.

The process of removing duplicates, or deduplication, can be implemented at different levels of specificity, from identifying exact matches to applying more nuanced, fuzzy matching techniques. Tools such as deduplicate-text-datasets and datasketch are effective in minimizing the redundancy in datasets by removing duplicate content. It is essential to acknowledge, as noted by researchers, that deduplication is resource-intensive, requiring significant computational power (both CPU and memory), especially given the substantial size of web-crawled datasets. Consequently, executing these operations in a distributed computing environment is often advisable to manage the demands efficiently.

Simple example:

```
import pandas as pd

# Example list of content with duplicates
content = ["Example text", "Unique content", "Example text", "Another unique piece", "Unique content"]

# Convert the list to a DataFrame
df = pd.DataFrame(content, columns=['Text'])

# Drop duplicates
df = df.drop_duplicates()

# Convert back to a list if needed
unique_content = df['Text'].tolist()

print(unique_content)
```

Output:

```
['Example text', 'Unique content', 'Another unique piece']
```

Data Decontamination

Ensuring the cleanliness and integrity of data in machine learning involves straightforward practices like separating training and testing datasets. Yet, for large language models (LLMs) drawing both training and evaluation data from the expansive terrain of the Internet, maintaining a clear distinction becomes a complex challenge.

For instance, the effectiveness of an LLM in understanding and responding to question-answer pairs during evaluation might be overstated if those very pairs were part of the model's training data. The practice of decontamination is crucial in this context, aiming to exclude any elements from the training data that are already part of benchmark datasets used for model evaluation. OpenAI, for example, intentionally omitted Wikipedia content from their WebText dataset training materials, given Wikipedia's extensive use in benchmark datasets. Similarly, EleutherAI introduced methods for benchmark dataset decontamination through their lm-eval harness package, accommodating scenarios where training data decontamination wasn't practical.

Decontamination addresses two main issues:

- **Input-and-output contamination**, where the model might simply replicate answers from its training data rather than generating insights, particularly relevant for tasks like abstractive summarization where the desired output might already exist within the training corpus.

- **Input contamination**, where the presence of evaluation examples, sans labels, in training data could artificially inflate the model's performance metrics in zero-shot or few-shot evaluations.

Simple example:

```
import pandas as pd

# Sample DataFrame with a mix of sensitive and non-sensitive rows
data = {'Text': ["User email is example@example.com",
                "Contact us at contact@example.net",
                "Our support email is support@example.org"],
        'IsSensitive': [True, True, False]}

df = pd.DataFrame(data)

# Remove sensitive rows directly without creating an intermediate slice
df_non_sensitive = df.loc[~df['IsSensitive']].copy()

# Now df_non_sensitive contains only non-sensitive rows, and we've avoided the warning
print(df_non_sensitive)
```

Output:

```
Text  IsSensitive
2  Our support email is support@example.org    False
```

The provided code performs the following actions:

- **Imports the Pandas Library:** This is a popular Python library used for data manipulation and analysis.

- **Creates a Sample DataFrame:** It uses a dictionary where keys represent column names ("'Text'" and "'IsSensitive'") and values are lists containing the rows for each column. The "'Text'" column contains strings, and the "'IsSensitive'" column contains Boolean values ("True" or "False") indicating whether the corresponding row is considered sensitive.

- **Filters Out Sensitive Rows:** The code uses "df.loc[~df['IsSensitive']]" to select only the rows where the "'IsSensitive'" column is "False" (i.e., nonsensitive rows). The tilde ("~") is a bitwise NOT operator used here to invert the Boolean series, thus selecting rows that are not marked as sensitive.

- **Creates a Copy of the Filtered DataFrame**: By calling ".copy()" on the filtered DataFrame, it ensures that "df_non_sensitive" is a new DataFrame independent of the original "df". This step is crucial to avoid any "SettingWithCopy" warnings from pandas, which can occur when modifying a DataFrame that is a slice of another DataFrame.

- **Prints the Nonsensitive DataFrame:** Finally, it prints "df_non_sensitive", which now contains only the rows from the original DataFrame that were marked as nonsensitive. The sensitive rows have been removed.

Addressing Toxicity and Bias

The vastness of web-sourced corpora inevitably includes a spectrum of content, with a notable presence of toxic and biased material. For example, studies like

RealToxicityPrompts[1] have quantified the prevalence of toxic content within widely used datasets, highlighting the necessity of filtering out such content to prevent perpetuating harmful biases in model outputs.

Techniques and tools like Perspective API serve to identify and mitigate the inclusion of toxic materials in training datasets, ensuring the resulting language models do not propagate or amplify these biases. Nevertheless, filtering for toxicity and bias demands meticulous consideration to avoid silencing marginalized voices or reinforcing dominant narratives, necessitating a comprehensive analysis of the content for pejorative language and biases related to gender, religion, and other sensitive areas before training.

Here's a simplified approach using Python:

- **Data Loading and Preprocessing:** Load the dataset and preprocess it for analysis. This typically involves cleaning the text (removing special characters, lowercasing, etc.).

- **Toxicity Detection:** Use a pre-trained model or API that specializes in detecting toxic content. Google's Perspective API is an example of a tool that can be used for this purpose.

- **Bias Detection:** Implement or use existing tools for detecting bias in text. This can involve checking for stereotypical language, underrepresentation of certain groups, etc.

- **Data Filtering and Annotation:** Based on toxicity and bias scores, filter out highly toxic or biased content, or annotate it for further review.

- **Review and Adjust:** Manually review a subset of flagged texts to ensure that the automated process is accurately identifying problematic content. Adjust the thresholds or methods as necessary.

- **Dataset Augmentation:** Optionally, augment the dataset with more diverse and balanced content to address underrepresentation issues.

- **Final Dataset Preparation:** Prepare the final, cleaned dataset for training by splitting it into training, validation, and test sets.

---

[1] Samuel Gehman, Suchin Gururangan, Maarten Sap, Yejin Choi, Noah A. Smith, REALTOXICITYPROMPTS: Evaluating Neural Toxic Degeneration in Language Models, Paul G. Allen School of Computer Science & Engineering, University of Washington, Allen Institute for Artificial Intelligence, Seattle, USA

CHAPTER 7   HARNESSING PYTHON 3.11 AND PYTHON LIBRARIES FOR LLM DEVELOPMENT

The following is a basic outline of a Python program that incorporates these steps. It assumes you have access to a toxicity detection API and functions for bias detection, which you may need to implement or integrate based on available resources:

Simple example:

```python
import pandas as pd
from your_toxicity_detection_tool import detect_toxicity
from your_bias_detection_tool import detect_bias

# Load dataset
def load_dataset(file_path):
    return pd.read_csv(file_path)

# Preprocess text
def preprocess_text(text):
    # Implement text cleaning here
    return text.lower()

# Detect and filter toxic content
def filter_toxic_content(data):
    data['toxicity_score'] = data['text'].apply(detect_toxicity)
    return data[data['toxicity_score'] < 0.5]  # Adjust threshold as needed

# Detect and annotate biased content
def annotate_biased_content(data):
    data['bias_score'] = data['text'].apply(detect_bias)
    data['is_biased'] = data['bias_score'] > 0.5  # Adjust threshold
    as needed
    return data

# Main function
def main():
    dataset_path = 'path_to_your_dataset.csv'
    dataset = load_dataset(dataset_path)
    dataset['text'] = dataset['text'].apply(preprocess_text)

    # Filter and annotate dataset
    dataset = filter_toxic_content(dataset)
    dataset = annotate_biased_content(dataset)
```

```
    # Optionally, review and adjust the dataset here

    # Prepare final dataset
    dataset.to_csv('cleaned_dataset.csv', index=False)
    print("Dataset cleaning and preparation completed.")
if __name__ == "__main__":
    main()
```

Protecting Personally Identifiable Information (PII)

The aggregation of large datasets also brings to the fore the critical issue of managing personally identifiable information (PII), encompassing details from names and social identification numbers to medical records. Legal and ethical standards necessitate the careful handling of PII, either through anonymization or outright removal, to safeguard privacy before using such data in language model training. Tools like presidio and pii-codex offer methodologies for detecting, analyzing, and managing PII, underscoring the importance of responsible data management practices in the development of language models.

Managing Missing Data

Addressing missing values in a dataset is crucial for maintaining the integrity of model training. Options for dealing with missing data include elimination or imputation. Removing rows or columns with missing values is a direct approach but reduces the volume of data available for training.

Alternatively, imputation techniques, which involve substituting missing values with estimated ones based on mean, median, or more sophisticated predictions like regression, can preserve data quantity. Advanced machine learning imputation methods, such as missForest and k-nearest neighbors, have been validated as effective through comparative studies.

CHAPTER 7 HARNESSING PYTHON 3.11 AND PYTHON LIBRARIES FOR LLM DEVELOPMENT

Simple example:

```python
import pandas as pd
from sklearn.impute import SimpleImputer

# Sample data creation
data = {
    'Feature1': [1, 2, None, 4],
    'Feature2': [None, 2, 3, 4],
    'Feature3': [1, None, 3, 4]
}
df = pd.DataFrame(data)

print("Original DataFrame:")
print(df)

# Handling missing data

## Option 1: Remove rows with missing data
df_dropped = df.dropna()
print("\nDataFrame after removing rows with missing data:")
print(df_dropped)

## Option 2: Impute missing values with mean
imputer = SimpleImputer(strategy='mean')
df_filled_mean = pd.DataFrame(imputer.fit_transform(df), columns=df.columns)
print("\nDataFrame after imputing missing values with mean:")
print(df_filled_mean)

## Option 3: Impute missing values with median
imputer.strategy = 'median'
df_filled_median = pd.DataFrame(imputer.fit_transform(df), columns=df.columns)
print("\nDataFrame after imputing missing values with median:")
print(df_filled_median)

## Option 4: Impute missing values with most frequent
imputer.strategy = 'most_frequent'
```

```
df_filled_most_frequent = pd.DataFrame(imputer.fit_transform(df),
columns=df.columns)
print("\nDataFrame after imputing missing values with most frequent
value:")
print(df_filled_most_frequent)
```

Output:

```
Original DataFrame:
   Feature1  Feature2  Feature3
0     1.0       NaN       1.0
1     2.0       2.0       NaN
2     NaN       3.0       3.0
3     4.0       4.0       4.0

DataFrame after removing rows with missing data:
   Feature1  Feature2  Feature3
3     4.0       4.0       4.0

DataFrame after imputing missing values with mean:
   Feature1  Feature2  Feature3
0  1.000000    3.0     1.000000
1  2.000000    2.0     2.666667
2  2.333333    3.0     3.000000
3  4.000000    4.0     4.000000

DataFrame after imputing missing values with median:
   Feature1  Feature2  Feature3
0     1.0       3.0       1.0
1     2.0       2.0       3.0
2     2.0       3.0       3.0
3     4.0       4.0       4.0

DataFrame after imputing missing values with most frequent value:
   Feature1  Feature2  Feature3
0     1.0       2.0       1.0
1     2.0       2.0       1.0
2     1.0       3.0       3.0
3     4.0       4.0       4.0
```

This program demonstrates four basic strategies for handling missing data:

- **Removing Rows with Missing Data:** This is the simplest approach where any row containing at least one null value is dropped.

- **Imputing Missing Values with Mean:** This replaces missing values with the mean value of each column. Suitable for numerical data without extreme outliers.

- **Imputing Missing Values with Median:** This method replaces missing values with the median value of each column. It's more robust to outliers than the mean.

- **Imputing Missing Values with the Most Frequent Value:** This replaces missing values with the most frequent value in each column. It's suitable for categorical data.

These methods are basic and widely used, but the choice of method depends on the dataset's specifics and the problem at hand. More advanced techniques, such as using models to predict missing values or employing deep learning for imputation, can also be explored for complex datasets.

Enhancing Datasets Through Augmentation

Data augmentation is a strategy to amplify the size and diversity of datasets, particularly valuable in scenarios where data is scarce. This technique is especially prevalent in training for machine translation and computer vision models due to its cost-effectiveness. By applying transformations like flipping, rotating, and scaling, it's possible to generate new, realistic variations of existing data, which is crucial in areas requiring extensive datasets, such as medical imaging. Deep AutoAugment, a recent advancement, has shown promise in enhancing data augmentation, demonstrating improved performance on benchmarks like ImageNet.

Data Normalization

Normalization plays a crucial role in standardizing the structure of dataset features to a consistent scale, enhancing the efficiency and accuracy of machine learning models. Techniques such as Min-Max scaling, log transformation, and z-score standardization are commonly employed by machine learning practitioners to achieve this uniformity.

CHAPTER 7　HARNESSING PYTHON 3.11 AND PYTHON LIBRARIES FOR LLM DEVELOPMENT

By adjusting the data to fit within a more restricted range, normalization facilitates quicker model convergence. Research in the field of data science has revealed that applying normalization techniques to datasets can enhance the performance of multiclass classification models by as much as 6%.

Simple example:

```
import pandas as pd
from sklearn.preprocessing import MinMaxScaler, StandardScaler

# Sample data creation
data = {
    'Feature1': [1, 2, 3, 4],
    'Feature2': [10, 20, 30, 40],
    'Feature3': [100, 200, 300, 400]
}
df = pd.DataFrame(data)

print("Original DataFrame:")
print(df)

# Data normalization

## Option 1: Min-Max Scaling
min_max_scaler = MinMaxScaler()
df_min_max_scaled = pd.DataFrame(min_max_scaler.fit_transform(df),
columns=df.columns)
print("\nDataFrame after Min-Max Scaling (normalized between 0 and 1):")
print(df_min_max_scaled)

## Option 2: Standardization (Z-score Normalization)
standard_scaler = StandardScaler()
df_standard_scaled = pd.DataFrame(standard_scaler.fit_transform(df),
columns=df.columns)
print("\nDataFrame after Standardization (Z-score normalization):")
print(df_standard_scaled)
```

Output:

```
Original DataFrame:
   Feature1  Feature2  Feature3
0         1        10       100
1         2        20       200
2         3        30       300
3         4        40       400

DataFrame after Min-Max Scaling (normalized between 0 and 1):
   Feature1  Feature2  Feature3
0  0.000000  0.000000  0.000000
1  0.333333  0.333333  0.333333
2  0.666667  0.666667  0.666667
3  1.000000  1.000000  1.000000

DataFrame after Standardization (Z-score normalization):
   Feature1  Feature2  Feature3
0 -1.341641 -1.341641 -1.341641
1 -0.447214 -0.447214 -0.447214
2  0.447214  0.447214  0.447214
3  1.341641  1.341641  1.341641
```

Explanation

- **Min-Max Scaling:** This technique scales and translates each feature individually such that it is in the given range on the training set, for example, between zero and one. This is useful when you need to bound the feature values strictly within two values.

- **Standardization (Z-score Normalization):** This technique standardizes features by removing the mean and scaling to unit variance. This is often more useful than Min-Max scaling for algorithms that assume all features are centered around zero and have the same variance.

These normalization methods are fundamental and widely applicable across various types of datasets. The choice between Min-Max scaling and Standardization depends on the specific requirements of your model and the characteristics of your data.

For instance, if your model needs input features within a specific range, Min-Max scaling might be more appropriate. On the other hand, if your model benefits from features having properties of a standard normal distribution (mean=0, variance=1), then Standardization would be the better choice.

Data Parsing

Parsing is the process of breaking down data to understand its syntax and extract useful information. This information then becomes input for large language models (LLM). In the realm of structured data, such as XML, JSON, or HTML, parsing is straightforward as it involves data formats with clear organization. For natural language processing (NLP), parsing takes on the task of deciphering the grammatical structure of sentences or phrases, which is essential for applications like machine translation, text summarization, and sentiment analysis.

Moreover, parsing extends to making sense of semi-structured or unstructured data sources, including email messages, social media content, or web pages. This capability is crucial for performing tasks like topic modeling, recognizing entities, and extracting relationships between them.

Simple example:

```
import string

def preprocess_text(file_path):
    """
    This function reads a text file and preprocesses it by:
    - Removing punctuation
    - Converting to lowercase
    - Splitting into words
    """
    # Define a translation table to remove punctuation
    translator = str.maketrans('', '', string.punctuation)

    # Read the file
    with open(file_path, 'r', encoding='utf-8') as file:
        text = file.read()
```

```python
    # Remove punctuation
    text = text.translate(translator)

    # Convert text to lowercase
    text = text.lower()

    # Split text into words
    words = text.split()

    return words

# Path to your data file
file_path = 'sample_data.txt'

# Preprocess the text
parsed_data = preprocess_text(file_path)

# Print the first 10 words to demonstrate the output
print(parsed_data[:10])
```

Output:

['lorem', 'ipsum', 'is', 'simply', 'dummy', 'text', 'of', 'the', 'printing', 'and']

Sample_data.txt content:

Lorem Ipsum is simply dummy text of the printing and typesetting industry. Lorem Ipsum has been the industry's standard dummy text ever since the 1500s, when an unknown printer took a galley of type and scrambled it to make a type specimen book. It has survived not only five centuries, but also the leap into electronic typesetting, remaining essentially unchanged. It was popularised in the 1960s with the release of Letraset sheets containing Lorem Ipsum passages, and more recently with desktop publishing software like Aldus PageMaker including versions of Lorem Ipsum.

Explanation

- The preprocess_text function takes a file path as input and returns a list of preprocessed words.
- It uses Python's built-in string module to remove punctuation and convert all text to lowercase for uniformity.

- The text is then split into individual words using the split() method.
- Finally, the script prints the first ten words from the processed dataset to show the result of the parsing.

This example is quite basic and intended for demonstration purposes. Depending on your specific needs, you might want to include additional preprocessing steps like removing stop words, stemming, lemmatization, or handling special text patterns and emojis.

Tokenization

Tokenization is the process of dividing text into smaller pieces, known as tokens. These tokens can range from individual words and subwords to characters. This segmentation turns complex text into a simpler, structured format that the model can efficiently process. By dissecting text into tokens, the model acquires a detailed insight into the nuances of language and its syntax, facilitating the generation and analysis of coherent word sequences.

Additionally, tokenization plays a critical role in establishing a vocabulary and developing word embeddings, which are essential for the model's ability to comprehend and produce language. This foundational step is vital for the preprocessing of text in large language models (LLMs), setting the stage for advanced language modeling and understanding.

Simple example:

```
import nltk
from nltk.tokenize import word_tokenize
nltk.download('punkt')

# Sample text
text = "Hello, world! This is an example of tokenization for language models."

# Perform word-level tokenization
tokens = word_tokenize(text)

print(tokens)
```

Output:

```
['Hello', ',', 'world', '!', 'This', 'is', 'an', 'example', 'of',
'tokenization', 'for', 'language', 'models', '.']
```

Explanation

- Install nltk with the command `pip install nltk`.

- First, the script imports nltk and the word_tokenize function from nltk.tokenize. The word_tokenize function is designed to split text into words using the Punkt tokenizer.

- A sample text is defined for the purpose of tokenization.

- The word_tokenize function is then called with the sample text as its argument, which returns a list of word tokens.

- Finally, it prints the list of tokens to show the result of the tokenization.

This example demonstrates basic word-level tokenization, which is suitable for many natural language processing (NLP) tasks. However, when working with LLMs, especially those using models like BERT or GPT, you might use more sophisticated tokenizers like byte pair encoding (BPE), WordPiece, or SentencePiece.

These tokenizers are capable of breaking text down into subword units, helping the model handle a wider variety of words, including those not seen during training, more efficiently. Many deep learning frameworks and libraries, such as Hugging Face's Transformers, provide easy access to these advanced tokenizers.

Stemming and Lemmatization

Stemming and lemmatization are key text preprocessing methods that aim to simplify words to their fundamental form, thereby reducing the complexity and size of the model's vocabulary.

Stemming is a basic technique that trims off the ends of words to capture their root forms, often leading to the removal of derivational affixes. Lemmatization, in contrast, is a more nuanced approach that takes into account the word's contextual usage and its grammatical category to accurately condense it to its lemma, or dictionary form. These strategies streamline text data, enhancing the model's efficiency in learning and understanding language.

CHAPTER 7 HARNESSING PYTHON 3.11 AND PYTHON LIBRARIES FOR LLM DEVELOPMENT

To do it, first you need to install NLTK with the command `pip install nltk`, then download the following data needed:

```
import nltk
nltk.download('wordnet')
nltk.download('omw-1.4')
nltk.download('punkt')
```

Example for stemming:

```
import nltk
from nltk.stem import PorterStemmer
from nltk.tokenize import word_tokenize

nltk.download('punkt')

# Initialize the stemmer
stemmer = PorterStemmer()

# Sample text
text = "The leaves on the tree are falling quickly due to the strong wind."

# Tokenize the text
tokens = word_tokenize(text)

# Stem each word in the text
stemmed_words = [stemmer.stem(word) for word in tokens]

print("Original Words:", tokens)
print("Stemmed Words:", stemmed_words)
```

Output:

```
Original Words: ['The', 'leaves', 'on', 'the', 'tree', 'are', 'falling', 'quickly', 'due', 'to', 'the', 'strong', 'wind', '.']
Stemmed Words: ['the', 'leav', 'on', 'the', 'tree', 'are', 'fall', 'quickli', 'due', 'to', 'the', 'strong', 'wind', '.']
```

Example for lemmatization:

```
import nltk
from nltk.stem import WordNetLemmatizer
from nltk.tokenize import word_tokenize
nltk.download('wordnet')

# Initialize the lemmatizer
lemmatizer = WordNetLemmatizer()

# Sample text
text = "The leaves on the tree were falling quickly due to the strong winds."

# Tokenize the text
tokens = word_tokenize(text)

# Lemmatize each word in the text
lemmatized_words = [lemmatizer.lemmatize(word) for word in tokens]

print("Original Words:", tokens)
print("Lemmatized Words:", lemmatized_words)
```

Output:

```
Original Words: ['The', 'leaves', 'on', 'the', 'tree', 'were', 'falling', 'quickly', 'due', 'to', 'the', 'strong', 'winds', '.']
Lemmatized Words: ['The', 'leaf', 'on', 'the', 'tree', 'were', 'falling', 'quickly', 'due', 'to', 'the', 'strong', 'wind', '.']
```

Explanation

- **Stemming:** The PorterStemmer class from NLTK is used for stemming, which reduces words to their root form by crudely chopping off the ends of words. This can sometimes lead to words that are not lexicographically correct.

- **Lemmatization:** The WordNetLemmatizer class requires WordNet Data, and it lemmatizes words by using a vocabulary and morphological analysis of words, returning the base or dictionary form of a word, which is known as the lemma.

Feature Engineering for Large Language Models

Feature creation is a pivotal process in machine learning that involves developing meaningful attributes or representations to facilitate the mapping of input data to desired outputs. For large language models (LLMs), this typically means creating embeddings – word or contextual – that adeptly capture the semantic and syntactic nuances of words within a multidimensional space. This capability enhances the model's understanding and generation of language.

Feature creation is a deliberate approach to boost model efficacy by injecting additional insights or organization into the input data. While data preprocessing readies the data for processing, feature creation refines it further, optimizing it for consumption by machine learning algorithms.

Word Embeddings

This process converts words or phrases into numerical vectors, positioning words with similar meanings close together in a continuous vector space. Static word embedding techniques like Word2Vec, GloVe, and fastText are renowned for producing these compact, multidimensional representations of text.

Word embeddings capture the essence of word context and semantic relationships, facilitating language models in tasks such as text classification, sentiment analysis, and language translation by understanding word usage and associations.

Contextual Embeddings

Offering a leap beyond traditional word embeddings, contextual embeddings generate dynamic word representations based on their usage in sentences. This approach, utilized by models like GPT and BERT, allows for the nuanced differentiation of meanings in polysemous words (words with several meanings) and homonyms (words identical in spelling but varying in meaning), like the word "bank," which could denote a financial establishment or a river's edge.

Contextual embeddings dynamically adjust word representations to reflect their specific context within sentences, significantly improving LLMs' performance across a spectrum of NLP applications by capturing the intricate variances in word meanings.

Subword Embeddings

Subword embeddings represent another innovative strategy, breaking down words into smaller subword units or vectors. This technique proves invaluable for managing rare or out-of-vocabulary (OOV) words, which fall outside the model's known vocabulary. By dissecting words into their subcomponents, the model can still attribute meaningful representations to these unfamiliar terms.

Techniques like byte pair encoding (BPE) and WordPiece are instrumental in this process. BPE progressively merges frequent subword pairs, whereas WordPiece divides words into characters before combining common character pairs. These methods adeptly grasp the morphological structures of words, boosting the model's capacity to handle a vast and varied vocabulary, thereby sharpening its semantic and syntactic discernment of words.

To make some embeddings, first you need to install transformers with the following command:

```
pip install transformers
```

Then write the following Python code:

```python
from transformers import BertTokenizer, BertModel
import torch

# Initialize the tokenizer and model
tokenizer = BertTokenizer.from_pretrained('bert-base-uncased')
model = BertModel.from_pretrained('bert-base-uncased')

# Encode text
text = "Hello, world!"
encoded_input = tokenizer(text, return_tensors='pt')

# Get embeddings
with torch.no_grad():
    outputs = model(**encoded_input)

# The last hidden state is the sequence of hidden states of the last layer
# of the model
last_hidden_states = outputs.last_hidden_state
```

```
# For simplicity, we can take the mean of the last hidden state as the
sentence embedding
sentence_embedding = torch.mean(last_hidden_states, dim=1)
print(sentence_embedding)
```

Output:

```
tensor([[-1.0990e-01, 8.5800e-02, 3.6918e-01, -3.1260e-01, 2.9934e-02,
1.0340e-01, 6.7850e-01, 6.0747e-01, -2.2346e-01, -5.2585e-01,
3.2231e-02, -5.4380e-01, -2.6335e-01, 5.3497e-01, -5.0481e-01,
....
2.8772e-02]])
```

This code snippet does the following:

- It loads the BERT tokenizer and model.
- It tokenizes the input text.
- It passes the tokenized input through the model to obtain the embeddings.
- Finally, it calculates a simple sentence embedding by taking the mean of the last hidden state output across the token dimension.

Keep in mind that this is a basic approach to obtaining sentence embeddings and there are more sophisticated methods for specific tasks or to capture deeper semantic meanings.

Best Practices for Data Processing

Effective management of training data for large language models (LLMs) is crucial for maximizing the efficiency and accuracy of these models. Adhering to best practices in data management not only preserves the quality and integrity of the data but also addresses common obstacles, facilitating a smoother data preparation process. Implementing these guidelines can significantly impact the performance of LLMs.

Implementing Strong Data Cleansing Protocols

Maintaining high standards of data cleanliness is crucial for the successful training of LLMs. This involves diligently cleaning data to eliminate inaccuracies, redundancies, or irrelevant details that could detrimentally affect model performance. Establishing clear protocols for data validation and cleansing ensures the training of LLMs on dependable, quality data, thereby enhancing model accuracy and robustness across applications.

Proactive Bias Management

Addressing bias in training data is essential for preventing outcomes that are unfair, unbalanced, or potentially biased. This encompasses both overt biases, such as prejudicial language, and more subtle biases, like the underrepresentation of specific groups or viewpoints. By actively curating data, selecting diverse datasets, and meticulously reviewing model outputs for bias, you can help ensure your LLMs function equitably and inclusively.

Implementing Continuous Quality Control and Feedback Mechanisms

To maintain the relevance and accuracy of LLMs, it's vital to have ongoing quality control and feedback mechanisms. Such systems facilitate the early detection of data discrepancies, model performance issues, or new biases, allowing for prompt corrective measures. This continuous improvement cycle, leveraging both performance data and user feedback, ensures the LLMs remain effective and current.

Fostering Interdisciplinary Collaboration

Encouraging collaboration across different expertise areas – uniting data engineers, machine learning specialists, and subject matter experts – can significantly enhance the data preparation and model training process. This collaborative environment ensures a holistic approach to data quality and model development, resulting in more sophisticated and accurate LLMs.

Prioritizing Educational Growth and Skill Development

Ensuring that teams involved in data management for LLMs are well versed in the latest methodologies, tools, and industry insights is key to maintaining high-quality data handling practices. Organizing ongoing education programs, such as workshops and seminars, and fostering a culture of continuous learning, empowers team members to enhance their expertise. This commitment to skill enhancement is crucial for navigating the complexities of LLM data management, leading to the development of more refined and precise models.

Delving into Key Libraries

In the realm of software development and data science, libraries are indispensable tools that enhance productivity, efficiency, and functionality. These libraries provide functions, allowing developers to implement complex tasks without starting from scratch. Delving into key libraries involves exploring essential frameworks and packages that streamline processes, from data manipulation and visualization to deep learning solutions. This exploration not only empowers developers with robust tools but also deepens their understanding of best practices and advanced techniques within the field of deep learning and large language models.

Natural Language Processing and Advanced Analytics

- **Transformers by Hugging Face:** This extensive library serves as a hub for cutting-edge NLP models, making advanced language processing accessible.

- **Gensim:** Focused on uncovering latent semantic patterns and topic modeling within vast text corpora, Gensim excels in understanding textual meaning.

- **TextBlob:** Designed for simplicity, TextBlob provides an intuitive API for tackling standard NLP operations, streamlining the manipulation of text.

- **Natural Language Toolkit (NLTK):** A foundational library offering a comprehensive toolkit for various NLP functions, renowned for its versatility.

- **Polyglot:** Optimized for handling NLP tasks across multiple languages, Polyglot brings a rich set of linguistic tools for global language processing.

- **Pattern:** A go-to for extracting data from the web, Pattern merges NLP, web scraping, and data mining functionalities for online content analysis.

Data Preparation and Integrity

- **NumPy:** A fundamental package for numerical computing in Python. It's widely used for data manipulation and supports operations on large, multidimensional arrays and matrices.

- **Pandas:** Offers data structures and operations for manipulating numerical tables and time series. It's ideal for data preprocessing tasks such as data cleaning, filtration, and aggregation.

- **Dask:** Provides parallel computing in Python using blocked algorithms and task scheduling. It's particularly useful for scaling Pandas workflows to handle larger-than-memory datasets on single machines by breaking the data into chunks.

- **TorchText:** Part of the PyTorch ecosystem, this library provides data processing utilities and popular datasets for the NLP domain. It's useful for building data pipelines for text processing and training.

- **TensorFlow Data Services (TFDS):** A collection of datasets ready to use with TensorFlow, complete with data loading and preprocessing capabilities. While not exclusively for NLP, it supports various datasets that can be used for training language models.

- **Scikit-learn:** While primarily a machine learning library, it offers a wide range of preprocessing functions for data normalization, scaling, and transformation. It can be used for feature extraction and normalization of textual data.

- **Apache Arrow:** A cross-language development platform for in-memory data. It provides efficient data interchange and processing capabilities, particularly useful for handling large datasets during preprocessing.

- **Unstructured:** Dedicated to refining unstructured data for ML algorithms, enhancing data readiness.

- **Pydantic:** Plays a crucial role in data validation and settings management within Python ecosystems.

- **Scrapy:** A powerful framework for web scraping, enabling the extraction of structured data from the Internet with ease.

Summary

This chapter focuses on harnessing Python 3.11 and Python libraries for the development of large language models (LLMs), introducing key frameworks such as LangChain and Hugging Face, and detailing their features, applications, and practical implementations.

As we conclude our journey through the fascinating world of Python and large language models (LLMs), it's evident how these powerful tools have revolutionized the landscape of technology and data science. Python's simplicity and versatility, coupled with the transformative capabilities of LLMs, offer unprecedented opportunities for innovation and problem-solving across various domains.

Whether you're developing sophisticated AI applications, automating complex workflows, or exploring new frontiers in natural language processing, the knowledge and skills you've gained from this book provide a strong foundation. As you continue to experiment, learn, and grow, remember that the fusion of creativity, curiosity, and technical prowess is the key to unlocking the full potential of Python and LLMs. The future is bright, and your contributions will undoubtedly shape the next wave of technological advancements.

Index

A

Adaptation
    alignment-tuning, 74
    alignment verification/utilization, 75
    fine-tuning, 74
    prompting, 76, 77
Approximate Nearest Neighbor (ANN), 327
Artificial intelligence (AI), 22
    education, 81
    Lamini.ai, 332
    LangChain, 303
    libraries, 303
    multimodal deep learning, 22
    natural language processing, 28
    OOP, 195, 196
    significant advancements, 90
    technologies, 1
Artificial neural networks (ANNs), 14
Attention-based language models
    attention mechanism, 20
    innovative developments, 20
    self-attention, 20
    transformer architecture, 21, 22
Attention mechanisms
    activation functions, 68
    cross/self-attention, 66
    definition, 66
    flash attention, 67
    full/sparse attention, 66
    layer normalization, 69
    data preprocessing, 71, 72
    deduplication, 72
    DeepNorm, 70
    distributed training, 70, 71
    LayerNorm, 69
    pre-norm/post-norm, 70
    layer normalization RMSNorm, 69
    positional encodings, 67, 68
    RoPE/Alibi, 68

B

Backpropagation through time (BPTT), 208
Bag-of-words (BOW) model, 9, 10, 12
Bengio, Yoshua, 13
Bidirectional Encoder Representations from Transformers (BERT), 21, 234
    architectural innovations, 242
    bidirectional strategy, 240
    classification token (CLS), 243
    definition, 240
    dual-phase strategy, 241
    functions, 241, 242
    input visualization, 242
    language processing, 243, 244
    pre-training/fine-tuning, 241
    transformer classification layer, 243
Business models
    client interactions, 97
    content moderation, 96

INDEX

Business models (*cont.*)
    emotions/sentiment analysis, 96
    high-quality digital content, 95
    recruitment, 98
    risk assessment/fraud prevention, 99
    sales cycle, 99
    SEO strategies, 95
    streamlining routine operations, 98
    translation, 97
    virtual teamwork, 98
Byte pair encoding (BPE), 65, 214, 216, 359, 363

C

Causal language modeling (CLM), 259
Chain-of-Thought (CoT), 76
Character Tokenization, 33
Chatbots/virtual assistants
    categories, 290
    concepts, 290
    consideration, 292
    customer support, 293
    data preprocessing/cleansing, 292
    document segmentation, 294
    embedding calculation, 294
    fine-tuning, 292
    GPT-3 techniques
        prompt template, 295
        text generation, 295–299
    integration/deployment, 292
    model selection, 292
    utilization, 290, 291
ChatGPT, 79
Chomsky, Noam, 5
Cohere model, 246, 247
    collaborations, 323
    command model, 324

    definition, 322
    embed models, 324
    Rerank model, 324–326
    transformative moment, 323
Content-Based Image Research (CBIR), 27
Contextual embeddings, 362
Convolutional neural networks (CNNs), 53

D

Deep learning (DL) techniques, 59
Deep learning model, 23

E

Embedding layers, 199
    components, 200
    feedforward layers, 205
        backpropagation, 206
        complex patterns hidden, 206, 207
        feedforward neural network, 205
        feedforward phase, 206
        hidden layer, 207
        input layers/hidden layers/output layers, 205
        neural network, 206
    matrix features, 201
    nodes, 202, 203
    recurrent layers, 208
        attention mechanism, 209
        BPTT, 208
        challenge/solution, 209
        components, 209
        cross-attention, 211
        global/local attention, 211
        masked attention, 211

multi-head attention, 210
self-attention, 210
sequential data processing, 208
sparse attention, 211
returning words, 203, 204
softmax layer, 204
vectors mirrors, 202
visualization, 201

F

Feedforward neural networks (FFNs), 199, 205

G

Gated Linear Unit (GLU), 68
Gated Recurrent Units (GRUs), 18, 60, 200
Gaussian Error Linear Unit (GELU), 68
Generative Pre-trained Transformer 4 (GPT-4)
characteristics, 235
ConceptARC, 237
foundational models, 235, 236
limitations, 236–240
performance comparison, 238
researchers, 240
specialized models, 237, 239
Generative question answering (GQA)
automated responses, 281
knowledge management, 281
real-world applications, 281
reports/unstructured documents, 281
Gigantic language models, 2
GPT4All project, 260, 261
Guanaco-65B, 257

H

Hallucination, LLM
benefits, 233
contexts, 232
data templates, 231
definition, 227
factually, 228
faithfulness, 228
high-quality training data, 231
human review, 232
implications, 229
mitigation risks, 229
overfitting/biases/inaccuracies, 228
primary types, 228
purpose and constraints, 231
restrict outcomes, 231
strategies, 230
testing/refinement, 231
Helpful, honest, and harmless (HHH), 75
Hidden Markov models (HMM), 23
Hugging Face model
code logs, 314
computer vision, 313
core components, 310
datasets library, 313–316
functionalities, 312
history, 310
Hub serves, 311
image processor, 315
Model Hub, 312
multimodal models, 313
operations, 313
pre-trained models, 310
segmentation mask, 315
technological community, 309
tokenizers, 313
transformers library, 311
version-controlled management, 311

INDEX

I

Information retrieval (IR) techniques, 8
Integrated development environment (IDE), 114
Inverse document frequency (IDF), 11–13

J, K

Jurassic-2 language model, 250

L

LaMDA, 257–259
Lamini.ai
   applications/use cases, 333–336
   definition, 332
   features/functionalities/advantages, 332, 333
   operational mechanics, 332
LangChain
   article summarization, 307–309
   capabilities, 306
   cloud storage services, 305
   components, 304
   definition, 303
   development process, 305, 306
   integrations, 305
Language modeling (LM), 59–62
Large Language Model Meta AI (LLaMA), 254, 255
Large language models (LLMs)
   accessibility, 85
   adaptation, 74–77
   advanced NLP, 80
   agents, 87
   architecture, 63, 64, 199, 234
   architecturesGPT-4, 235, 236
   attention, 65–72
   attention-based language, 20–22
   benefits, 78
   BERT, 240–244
   business models, 95–99
   capabilities, 22
   chatbots/virtual assistants, 290–298
   Chomsky's theory, 5
   computer vision/speech recognition, 1
   content creation, 82
   creative arts, 83
   disaster/management, 84
   DOCTOR, 7
   education, 81
   ELIZA program, 6–8
   embedding layers, 200
   engineering disciplines, 85, 86
   engineering feature, 362–364
   environmental science, 85
   ethics/bias/misinformation, 83
   evolutionary steps, 2
   factors, 22
   finance sector, 82, 83
   forecasting/finance, 84
   gaming industry, 84
   hallucination, 227–233
   healthcare communication/management, 80
   implications, 233, 234
   in-depth architecture, 64, 65
   industry analyses, 90
   key advantage, 91–94
   Lamini.ai, 332–336
   language modeling, 59–62
   legal/compliance assistance, 83
   libraries, 366, 367
   limitations, 87

adversarial attacks, 90
bias, 88
information hallucination, 89
localization, 82
Markov, Andrey/Shannon, Claude, 3, 4
medical applications, 79, 80
multimodal learning, 22, 233 (*see also*
   Multimodal deep learning)
neural language (*see* Neural language
   models (NLMs))
NLP (*see* Natural language
   processing (NLP))
personalized marketing/customer
   insights, 84
preparation/integrity, 367
processing data, 364
   bias management, 365
   cleansing protocols, 365
   educational growth/skill
      development, 366
   interdisciplinary collaboration, 365
   quality control/feedback
      mechanisms, 365
prompting, 299, 300
QA (*see* Question answering (QA))
research/data analysis, 82
rule-based approach, 6
simplified process, 62, 63
statistical language (*see* Statistical
   language processing)
statistical/rule-based models, 1
substantial attention, 1
sustainability, 85
tokenization, 65, 212–225, 358
training process, 77
transformer architectures, 72
   causal decoder, 73
   encoder-decoder, 73
   prefix decoder, 73
   pre-training objectives, 73
   translation (*see* Translation systems)
   translation services, 82
   unigramLM, 65
   versatile tools, 78
Latent Dirichlet Allocation (LDA), 42
Long short-term memory (LSTM), 14, 16,
   17, 21, 60, 200
Low Rank Adapters (LoRA), 257

M

Machine learning (ML), 337
   Hugging Face, 309
   translation system, 268
Markov, Andrey, 3, 4
Masked Language Model (MLM), 242
Mikolov, Tomas, 14
Multilayer Perceptron (MLP), 252
Multimodal deep learning
   alignment, 25
   applications, 26
   captioning image, 27
   co-learning, 25
   effective representations, 24
   emotion recognition, 27
   fundamental challenges, 23
   fusion, 24
   integral components, 26
   modalities, 22
   primary modalities, 23
   retrieval image, 27
   text-to-image generation, 27
   translation entails, 25
   unimodal networks, 26
Multimodal learning, data
   types, 57

INDEX

N

Named entity recognition (NER), 31, 36, 41, 243
Natural language generation (NLG), 53
Natural language processing (NLP), 5, 52, 80, 82, 312
   ambiguity/context/precision, 53
   attention-based language models, 20
   computational linguistics, 28
   computer programs, 28
   conventional techniques/LLMs, 55–57
   coreference resolution, 32
   data preprocessing, 29
   discourse, 30
   foundational principles, 54
   Hugging Face, 309
   human communication/computer comprehension, 28
   humans possess, 29
   interface, 28
   LangChain, 304
   language modelling, 60
   libraries, 366
   machine learning/deep learning models, 53
   NLG capabilities, 53
   NLU model, 52
   OOP, 196, 197
   parsing, 356
   part-of-speech tagging, 31
   pragmatics, 30
   primary components, 29
   real-world language, 54
   semantics, 30
   sentiment analysis, 32, 47–52
   speech recognition, 31
   statistical model, 53
   summarization, 273
   supervised methods, 52
   syntax encompasses, 30
   tasks, 31, 32
   text generation, 264
   text preprocessing/engineering, 32–43
   tokenization, 215
   tone/inflection/semantic analysis, 54
   unsupervised method, 52
   word embedding, 43–55
   word sense disambiguation, 31
Natural Language Toolkit (NLTK), 366
   extraction techniques, 41
   lemmatization, 39
   libraries, 38
   named entity recognition, 41
   n-grams, 40
   POS tagging, 40
   Python, 38
   stemming, 39
   stop word removal, 39
   TF-IDF, 41
   tokenization, 38
Natural language understanding (NLU), 52
Neural language models (NLMs), 60
   conditional language models, 19
   conventional n-gram model, 13
   distributed representation, 14
   encoder-decoder architectures, 18
   GRUs, 18
   LSTM model, 16, 17
   recurrent neural networks, 14, 15
   seq2seq models, 18, 19
   Word2Vec model, 14
Next Sentence Prediction (NSP), 242
Non-negative matrix factorization (NMF), 42
NormalFloat (NF4), 257

O

Object-Oriented Programming (OOP)
  abstraction, 169, 170
  AI, 195, 196
  attributes/methods, 167
  coding approach, 165
  core principle, 167
  encapsulation, 173, 174
  file handling
    append() mode, 185
    capabilities, 180
    End of Line (EOL), 180
    reading mode, 182–184
    reading/writing, 180–182
    split() function, 184
    write() function, 185
  getter and setter methods, 173
  inheritance, 170, 171
  modules
    asterisk (*) symbol, 177
    attributes, 176
    built-in modules, 178, 179
    creation, 174
    definitions/statements, 174
    dot (.) operator, 175
    "from" statement, 176
    import functions, 175
    locations, 177, 178
    rename, 178
    structure, 175
  NLP, 196, 197
  objects, 166
  polymorphism, 171, 172
  Python 3.11
    arbitrary literal string type, 193
    error messages, 189
    exception notes, 190, 191
    features, 186
    negative zero formatting, 194
    Required()/NotRequired(), 187, 188
    self type, 188, 189
    TOML files, 191–193
    TypedDict, 186, 187
    type variables, 193
    variadic generics, 194
  Python object
    attributes/methods, 168
    creation (Book object), 168
  structured programming, 166
OpenAI API
  capabilities, 316
  code completion service, 319
  customization, 317
  features, 317
  image generation, 319
  industries, 318–320
  pre-trained models, 317
  scalable infrastructure, 318
  sentiment analysis, 318
  summarized biography, 320–322
  text comparison, 318
  user-friendly interface, 318
Optical Character Recognition (OCR), 344
Orca 2, 258
Out-of-vocabulary (OOV), 363

P

Palmyra Base (5b), 259
Part-of-speech (POS), 43
Pathway interaction and collaboration
  Claude v1, 251, 252
  Falcon 40B, 252
    accessibility, 254
    design, 252

INDEX

Pathway interaction and
    collaboration (*cont.*)
  instruct variants, 254
  multi-query attention, 253
  training data, 253
 J2 platform, 250
 outputs, 250
 selective engagement, 249
Pathways Language Model 2 (PaLM 2)
 adaptive computational allocation, 249
 data acquisition/preparation, 248
 definition, 247
 foundational technology/
    components, 248
 independence, 249
 pathways architecture, 249
 pre-training phase/fine-tuning, 248
 transformer architecture, 248
Personally identifiable information
    (PII), 64, 350
Pinecone
 advantages, 328
 features, 328, 329
 innovative databases, 327
 numerous sectors, 329–331
 operational mechanics, 327
 vector databases, 327, 328
Pre-trained Language Models (PLMs), 60
Prompt erasure, 258
Prompting application
 benefits, 299
 code generation/translation, 300
 creative writing, 300
 fundamental model, 299
 template, 300
Python programming language,
    38–43, 101
 acquisition, 338

arguments
 anonymous, 163
 default values, 162
 keyword, 162
 positional arguments, 161
 variable-length, 162
augmentation, 353
blank lines/multiline statement, 108
Booleans
 arithmetic, 123
 built-in data types, 121
 integers/floating-point
    numbers, 122
 operators, 122, 123
comments, 107, 108
compact and readable programs, 102
conditionals/loops
 conditionals, 132, 133
 control flow, 132
 decision-making statements, 132
 else/elif clauses, 135, 136
 grouping statements, 133
 nested blocks, 134
 one-line if statements, 136
datasets, 337
data structures
 built-in data, 142
 data access/modification, 143
 dictionaries, 148–152
 lists, 144–147
 lists/dictionaries/tuples/sets, 143
 manipulation/inquiry, 147, 148
 stacks/queues/trees/linked
    lists, 143
data types
 built-in functions, 118
 complex numbers, 117
 delimiters/characteristics, 118

INDEX

floating-point, 117
handling special characters, 119
integers, 116
numeric types, 116
RAW string, 120
single/double quotes, 118
strings, 117, 118
triple-quoted strings, 120
decontamination, 346–348
dictionaries
    accessing elements, 150
    creation, 148
    functions, 151
    key-value pair, 149
    modification, 148
exception handling mechanism, 102
extensive datasets, 337
functions
    arguments, 161–164
    call function, 160, 161
    categories, 158
    method *vs.* functions, 160
    print/return, 160
    return statement, 159
    UDFs, 158
gathering/preparation, 337
handling document length, 343
high readability, 103
identifiers, 104
IDLE program, 114
indentation, 105
interactive/scripting mode, 114
justext and trafilatura
    methods, 340–344
lambda functions, 163, 164
loops, 136
    continue, break, and pass, 142
    control statements, 142

else statement, 138
infinite loop, 138
for loops, 139, 140
nested loops, 140, 141
outer loop/inner loop, 141
while, 137
macOS, 112, 113
Min-Max scaling/standardization, 355
missing values, 350–353
multi-line statements, 106
naming variables, 115
normalization methods, 353–356
OOP (*see* Object-Oriented Programming (OOP))
operating systems, 101, 108
operators, 123
    arithmetic operators, 124, 125
    assignment, 128–130
    bitwise, 127
    comparison, 125
    float/floor division, 124
    identity, 130
    logical operators, 126
    membership, 131
    ternary, 131
parsing, 356–358
PII models, 350
preprocessing data, 338
quotations, 106, 107
removing duplicate content, 344, 345
reserved words, 104
sentiment analysis, 47–52
sets, 155
    add elements, 155
    creation, 155
    difference() function, 157
    intersection() function, 157
    operations, 156, 157

Python programming language (*cont.*)
    symmetric_difference()
        function, 157
    union() function, 157
  standard modules, 102
  stemming/lemmatization, 359–361
  syntax/semantics, 102
  text files, 101
  text produced machines, 344
  tokenization, 358, 359
  toxicity/bias, 347–350
  training language models, 339
  tuples, 152–155
  Ubuntu/Debian and Fedora, 113
  variables serve, 114–116
  Windows, 109–112
  Zen, 103, 104
  Z-score normalization, 355

Q

Question answering (QA)
  chain setup, 286
  chatbot harnesses, 282, 283
  components, 283–287
  document analysis, 279
  embeddings/indexing, 280
  generative model, 280
  generative system, 279
  GQA (*see* Generative question answering (GQA))
  metadata dictionary, 286
  parsing/preparation documents, 280
  query processing/context retrieval, 280
  source code, 289–291
  systematic approach, 280

R

Rectified Linear Unit (ReLU), 68
Recurrent neural networks (RNNs), 14, 15, 24, 53, 60, 263
Reinforcement learning (RL), 75
Reinforcement Learning with Human Feedback (RLHF), 75
Restricted Boltzmann Machines (RBM), 23
Reward modeling (RM), 75

S

Search engine optimization (SEO), 95
Sentiment analysis
  categories, 48
  feature extraction, 48
  fundamental building blocks, 48, 49
  opinion mining, 47
  text classification
    advantages, 49
    applications, 50–52
    dynamic/influential domain, 49
    faces challenges, 50
    mechanics, 50
    supervised/unsupervised, 50
  text processing, 48
sequence-to-sequence (seq2seq) models, 18, 19
Shannon, Claude, 4, 5
Short-Term Memory, 17
StableLM, 258, 259
Statistical language models (SLMs), 60
Statistical language processing
  bag-of-words (BOW), 9, 10
  n-gram models, 8, 9
  noteworthy application, 9

text embedding, 9
TF-IDF, 10–12
Stochastic gradient descent (SGD), 77
Subword embeddings, 363, 364
Subword Tokenization, 33

T

Technology Innovation Institute (TII), 252
TensorFlow Data Services (TFDS), 367
Term frequency-inverse document frequency (TF-IDF), 8, 11–13
Text generation, LLM
    advantages, 264
    application, 264
    conceptualization/brainstorming, 265
    creative writing, 265
    dialogue crafting/characterization, 266
    drafting phase, 265
    human writing, 263
    poetry/experimental writing, 266
    significance, 264
    source code, 266–268
    transformer models, 263
    world-building elements, 266
Text preprocessing/engineering
    feature extraction, 37
    Python/NLTK, 38–43
    tokenization
        fundamental objective, 32
        lemmatization, 34
        methods, 32, 33
        morphemes, 35
        morphological segmentation, 35
        named entity recognition, 36
        parsing, 34
        pattern recognition, 32
        sentence breaking, 35
        sequential stages, 36
        stemming, 35
        word segmentation, 34
        word sense disambiguation, 34
Text summarization/document methods
    article application, 274–278
    definition, 273
    direct/abstractive/extractive, 273
Text-to-Text Transfer Transformer (T5)
    multitask learning approach, 246
    tasks/outputs, 245
    training/fine-tuning, 245
    transformer architecture, 246
Tokenization
    advantages, 213
    case sensitivity, 214
    characteristics, 215
    definition, 212
    distribution, 216, 217
    embedding visualization, 218
    encoder-decoder feedforward network, 220
    grasping contextual nuances/ambiguity, 214
    idiomatic expressions, 215
    limitations/challenges, 213
    model approaches, 214
    numeric data handling, 214
    positional embeddings, 219
    self-attention, 219, 220
    sequence-to-sequence translation, 221
    straightforward principle, 218
    trailing whitespaces, 214
    transformer model architecture, 221
    unique symbols/punctuation/characters, 215
    voluminous/complex, 212

INDEX

Tom's Obvious Minimal Language
(TOML), 191–193
Translation systems
  advantages, 268
  benefits, 269
  bias/cost/regulatory
    considerations, 269
  data collection and preprocessing, 269
  definition, 268
  enhanced efficiency, 270
  Google T5 model, 270–273
  pioneering opportunities, 270
  quality, 270
Tree of Thought (ToT), 76

U, V

Unimodal/Monomodal models, 23
User-defined functions (UDFs), 158

W, X, Y

Word embedding
  benefits, 44
  deep learning models, 44
  limitations, 45
  semantic analysis
    components, 46
    human language
      comprehension, 46
    key elements, 47
  techniques, 44
  vectors, 43
  words/documents, 43
  words/phrases, 362
Word Tokenization, 33

Z

Zero-shot/few-shot learning
  approaches, 225
  categories, 223
  data learning, 225
  definition, 222
  development models, 223
  few-shot learning, 226
    definition, 222
    real-world application, 222
    significance, 222
  one-shot learning, 225, 226
  significance, 224
  transformative approach, 224

Printed in the United States
by Baker & Taylor Publisher Services